Books are to be returned on or before
the last date below.

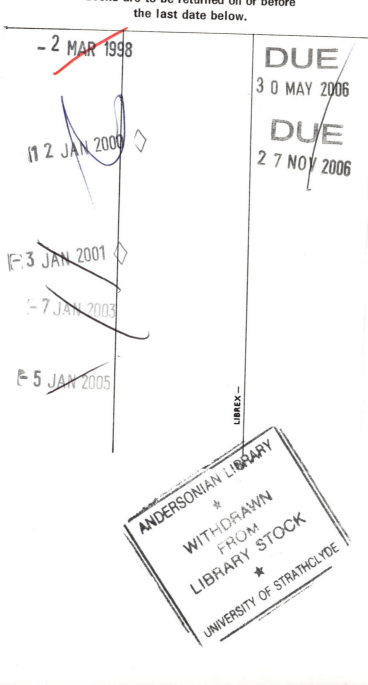

- 2 MAR 1998

DUE
3 0 MAY 2006

12 JAN 2000

DUE
2 7 NOV 2006

-3 JAN 2001

- 7 JAN 2003

- 5 JAN 2005

LIBREX—

Dehydration of Foods

Dehydration of Foods

SERIES EDITOR

Gustavo V. Barbosa-Cánovas, Washington State University

EDITORIAL BOARD

DEHYDRATION OF FOODS

GUSTAVO V. BARBOSA-CÁNOVAS
WASHINGTON STATE UNIVERSITY

HUMBERTO VEGA-MERCADO
MERCK SHARP & DOHME OUÍMICA DE PUERTO RICO

CHAPMAN & HALL

I(T)P® International Thomson Publishing

New York • Albany • Bonn • Boston • Cincinnati • Detroit • London • Madrid • Melbourne
Mexico City • Pacific Grove • Paris • San Francisco • Singapore • Tokyo • Toronto • Washington

Cover design: Trudi Gershenov

Printed in the United States of America

For more information, contact:

Chapman & Hall
115 Fifth Avenue
New York, NY 10003

Thomas Nelson Australia
102 Dodds Street
South Melbourne, 3205
Victoria, Australia

Nelson Canada
1120 Birchmount Road
Scarborough, Ontario
Canada M1K 5G4

International Thomson Editores
Campos Eliseos 385, Piso 7
Col. Polanco
11560 Mexico D. F.
Mexico

Chapman & Hall
2-6 Boundary Row
London SE1 8HN
England

Chapman & Hall GmbH
Postfach 100 263
D-69442 Weinheim
Germany

International Thomson
Publishing Asia
221 Henderson Road #05-10
Henderson Building
Singapore 0315

International Thomson
Publishing - Japan
Hirakawacho-cho Kyowa
Building, 3F
1-2-1 Hirakawacho-cho
Chiyoda-ku, 102 Tokyo
Japan

1 2 3 4 5 6 7 8 9 10 XXX 01 00 99 98 97 96

Library of Congress Cataloging-in-Publication Data

Barbosa-Cánovas, Gustavo V.
    Dehydration of foods / Gustavo V. Barbosa-Cánovas and Humberto
Vega-Mercado.
        p.   cm.
    Includes bibliographical references and index.
    ISBN 0-412-06421-9 (alk. paper)
    1. Food--Drying.   I. Vega-Mercado, Humberto.   II. Title.
    TP371.5.B365   1996   95-44230
    668' .0284--dc20   CIP

To order this or any other Chapman & Hall book, please contact **International Thomson Publishing, 7625 Empire Drive, Florence, KY 41042.** Phone: (606) 525-6600 or 1-800-842-3636. Fax: (606) 525-7778. e-mail: order@chaphall.com.

For a complete listing of Chapman & Hall's titles, send your request to **Chapman & Hall, Dept. BC, 115 Fifth Avenue, New York, NY 10003.**

JOIN US ON THE INTERNET
WWW: http://www.thomson.com
EMAIL: findit@kiosk.thomson.com

thomson.com is the on-line portal for the products, services and resources available from International Thomson Publishing (ITP).This Internet kiosk gives users immediate access to more than 34 ITP publishers and over 20,000 products.Through *thomson.com* Internet users can search catalogs, examine subject-specific resource centers and subscribe to electronic discussion lists.You can purchase ITP products from your local bookseller, or directly through *thomson.com*.

Visit Chapman & Hall's Internet Resource Center for information on our new publications, links of useful sites on the World Wide Web and an opportunity to join our e-mail mailing list. Point your browser to: http://www.chaphall.com/chaphall.html or http://www.chaphall.com/chaphall/foodsci.html for Food Science

A service of I(T)P

To our families

CONTENTS

APPENDIXES

PREFACE

Food dehydration is one of the most relevant and challenging unit operations in food processing, as well as a topic of continuous interest in food research. Information on fundamental aspects and applications of food dehydration has been disseminated mainly through research and review articles, edited books, trade magazines reports, and symposia presentations. It is apparent that not many books have been written to cover, in a systematic manner, basic and applied engineering aspects of this unit operation, as well as key issues dealing with physical and biochemical changes occurring during dehydration. This book was designed and developed as a useful reference for those individuals, either in the food industry or in academia, interested in an organized and updated review, from an engineering perspective, of the most important aspects of food dehydration. The book, comprising nine chapters, includes several tables, figures, diagrams, photographs, and extensive literature citation, as well as numerical examples, to facilitate the understanding of each module. A significant effort was made to identify, from a broad spectrum of references, those areas in food dehydration that are of current interest as well as those essential to a better understanding of the process in general. Some of the fundamental aspects covered in the book are: air–water mixtures; theory, prediction, and measurement of water activity in foods; glass transition as applied to food dehydration; the most accepted theories in food dehydration; and ideal dryers with and without recirculation. The most popular techniques in food dehydration, such as spray drying, freeze dehydration, tray drying, cabinet drying, and osmotic dehydration are extensively covered. Other approaches, although concisely described, offer the opportunity to appreciate the many options available to effectively dehydrate food products. The last chapter includes an extensive section dealing with the proper selection of packaging materials for dehydrated foods.

We sincerely hope this book will be a valuable addition to the food literature and will promote additional interest in food dehydration research, development, and implementation.

Gustavo V. Barbosa-Cánovas
and Humberto Vega-Mercado

CHAPTER 1

INTRODUCTION TO DEHYDRATION OF FOOD

1.0 INTRODUCTION

It is not known when the preservation of foods by dehydration began, but history does show that our ancestors learned how to dry foods by trial and error. Food dehydration eventually evolved within a scientific based environment and made possible the establishment of a worldwide industry, capable of providing a convenient and nutritious food supply.

The first record of drying is for vegetables, and appeared in the 18th century (Van Arsdel and Copley, 1963). Thereafter, the development of the drying industry was closely related to war scenarios around the world. British troops in the Crimea War (1854–1856) received dried vegetables from their homeland, Canadian dried vegetables were shipped to South Africa during the Boer War (1899–1902), and around 4500 tons of dehydrated vegetables were shipped from the United States during World War I. By 1919, among the products processed in the United States were green beans, cabbage, carrots, celery, potatoes, spinach, sweet corn, turnips, and soup mixtures.

Fruit dehydration in the United States made a significant turn at the end of the 1800s and the beginning of the 1900s with the development of artificial dryers to replace sun drying. Drum drying and spray drying were investigated and developed before the Second World War, and used extensively for milk products and eggs.

Dehydration is especially suitable for military purposes because of the space and weight saving capabilities it affords. In the United States the production of dehydrated food evolved from processes developed for wartime uses to a vigorously growing branch of the entire food processing industry. Table 1.1 summarizes dried food production in the United States during 1992 and 1993.

Drying is a process in which water is removed to halt or slow down the growth of spoilage microorganisms as well as the occurrence of chemical reactions. The terms dried and dehydrated are not synonymous. The US Department of Agriculture lists dehydrated food as those with no more than 2.5% water (dry basis) while dried food apply to any food product that has been exposed to

Table 1.1   Summary of dried food production for 1992 and 1993.

Product	1992	1993	units
Nonfat dry milk	872	948	million pounds
Dry whole milk	168	153	million pounds
Dry whey	1237	1196	million pounds
Dry edible beans	22	22	million cwt
Fruits	619	557	million pounds

From USDA (1994).

a water removal process which has more than 2.5% water (dry basis).

In addition to preservation, drying is used to reduce the cost or difficulty of packaging, handling, storing, and transporting by converting the raw food to a dry solid. This reduces the weight and sometimes the volume.

1.1 THEORETICAL ASPECTS

Theoretical concepts such as moisture content, dry and wet bulb temperature, relative humidity, humid heat, dew point, saturation condition, adiabatic saturation, water and enzymatic activity, microbial spoilage, crispness, viscosity, hardness, aroma, flavor, palatability, drying periods, and theories on mass transfer phenomena are discussed from the dehydration perspective in Chapters 2, 3, and 4.

The removal of water from a food is achieved mainly by use of dry air (except for some unit operations such as freeze drying and osmotic dehydration) which picks up the water from the surface of the product and carries it away. The engineering concepts used to explain the water–air relationship are presented in Chapter 2. Also included is an introduction to the design of an ideal dryer and mass and heat balances.

The process of drying foods not only affects the water content of the product, but other physical and chemical characteristics as well. Among the characteristics used to describe dried foods are water activity; sorption isotherms; microbial spoilage; enzymatic and nonenzymatic reactions; physical and structural phenomena; and destruction of nutrients, aroma, and flavors. Chapter 3 presents a comprehensive discussion of these characteristics and their implications in food dehydration.

What mechanisms are involved in the movement of water during the dehydration process? They can be summarized as water movement under capillary forces, diffusion of liquid due to concentration gradients, surface diffusion, water vapor diffusion in air-filled pores, flow due to pressure gradients, and flow due to a vaporization-condensation sequence. These mechanisms are covered in Chapter 4 with special attention to the falling rate period during which the physical and chemical characteristics of foods affect water removal during dehydration.

1.2 DRYERS

Heat required for drying may be supplied by convection, conduction, and radiation. Both direct and indirect drying are commonly found in food processing. Indirect drying systems are simpler both in concept and required equipment (Cook and DuMont, 1991). In indirect dryers, heat is conducted into the food by hot metal walls of the containing vessel and within food particles by direct contact of hot with cold particles. Direct dryers use hot gas, commonly air, which passes over or through the food. Heating is thus further improved and made more uniform than in indirect dryers.

The selection of a dryer needs to be based on the entire manufacturing process. Raw materials, intermediate product, and final product specifications and characteristics (i.e., final moisture content) need to be clearly defined. Preprocessing steps may be considered to remove water content prior to the final drying step (i.e., osmotic dehydration prior to freeze drying). The final assessment for the selection of a dryer should include, but not be limited to, the production capacity, initial moisture content of the product, particle size distribution, drying characteristics of the product, maximum allowable product temperature, explosion characteristics (i.e., spray or fluid bed dryings), moisture isotherms, and physical data of the mater-

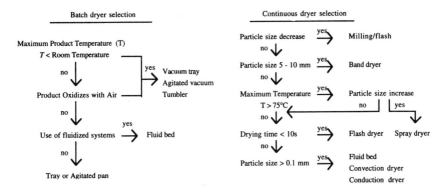

Figure 1.1   A possible approach to select a batch or continuous dryer.

ial. Figure 1.1 summarizes a possible procedure for the selection of either a batch or a continuous dryer.

Food dehydration is not limited to the selection of a dryer. The physicochemical concepts associated with food dehydration need to be understood for an appropriate assessment of the drying phenomena in any food product. Water activity, glass transition temperature, dehydration mechanisms and theories, and chemical and physical changes should be recognized as key elements for any food dehydration operation.

Cabinet and bed dryers are among the most commonly found in dehydration operations. Grains, fruits, and vegetables are processed using cabinets or bed dryers such as a kiln, tray, truck tray, double cone, rotary, flow conveyor, or tunnel. The dehydration operation may be batch, semicontinuous, or continuous depending on the type or amount of product and frequency of the drying operation. The use of steam dryers and heat pumps allows better retention of color in agricultural commodities such as cabbage and hay as well as increased heat energy efficiency compared to cabinet or bed dryers.

Spray drying is oriented to liquid foods with relatively high solids content. It involves the atomization of the feed into the drying medium and the recovery of the dried particles once the water is removed. The advantages of spray drying include constant specifications throughout the dryer, providing drying conditions are held constant, continuous and easy operation, and a diversity of designs available for processing both heat-sensitive and heat-resistant materials (Masters, 1991). Some examples of food products processed by spray drying are milk, eggs, coffee, tea, enzymes, and whey proteins.

Heat-sensitive foods can be processed by freeze dehydration, which was developed to overcome the loss of compounds responsible for flavor and aroma during conventional drying operations (Karel, 1975). Some food products that are commercially freeze dried include: extracts (coffee and tea), vegetables, fruits, meats, and fish (Schwartzberg, 1982; Dalgleish, 1990). Freeze-dried products are light in weight (10% to 15% original weight) and do not require refrigeration. Moisture levels as low as 2% are reached with freeze drying. Steaks, fish, and chicken can be dried without crushing or shredding the product.

Osmotic dehydration, in which water is removed by osmosis, is another process used to preserve food products. The water lost during osmotic dehydration can be divided into a period with a high rate of water removal followed by one with a decreasing rate of water removal. Temperature and concentration of the osmotic solution affect the rate of water loss. Osmotic dehydration is quicker than air or freeze drying because the removal of water occurs without a phase change (Farkas and Lazar, 1969; Raoult-Wack et al., 1989; Jayaraman and Das Gupta, 1992). Osmotic dehydration is used in the development of shelf-stable foods and is used as an effective hurdle against microbial growth by reducing the water activity of the food. It also offers a milder treatment than hot air dehydration. This safer and more efficient technology has been used in the preservation of fruits (i.e., apple, mango, banana, guava, papaya, cherry, passion fruit, tamarind, and pineapple); vegetables (i.e., pumpkin, tomato, okra, aubergin, pepper, carrot, potato); fish; and meat products.

In addition to the previously mentioned dehydration methods (i.e., cabinet dryers, spray drying, freeze dehydration, and osmotic dehydration), there are other dehydration alternatives used for food preservation: sun, vacuum-high temperature, drums, microwave, extrusion, fluidized bed, and pneumatic drying are just a few. Each one has specific applications and should be considered based on product characteristics and the final quality desired for the dried food. Also, the appropriate postdrying handling, packaging, and storage should be considered because light, storage atmosphere, and temperature affect the stability of dried foods.

1.3 FINAL REMARKS

Food dehydration is not a trivial unit operation. Several factors affect the performance of a dryer as well as the quality of the product during the drying operation. The physical and chemical changes during a drying operation will improve certain characteristics of the intended products but will also decrease the amount of nutrients and organoleptic properties. With proper handling, however, these reactions and physical changes will ensure a highly nutritional food product with a significantly extended shelf-life.

1.4 REFERENCES

Cook, E. M. and DuMont, H. D. 1991. *Process Drying Practice*. McGraw-Hill, New York.

Dalgleish, J. McN. 1990. *Freeze-Drying for the Food Industries.* Elsevier Applied Science, New York.

Farkas, D. F. and Lazar, M. E. 1969. Osmotic dehydration of apple pieces. Effect of temperature and syrup concentration on rates. *Food Technol.* 23:688–690.

Jayaraman, K. S. and Das Gupta, D. K. 1992. Dehydration of fruit and vegetables—recent developments in principles and techniques. *Drying Technol.* 10:1–50.

Karel, M. 1975. Freeze dehydration of foods. In *Principles of Food Science. Part II: Physical Principles of Food Preservation,* edited by M. Karel, O. W. Fennema, and D. B. Lund. Marcel Dekker, New York.

Masters, K. 1991. *Spray Drying Handbook.* Fifth edition. John Wiley & Sons, New York.

Raoult-Wack, A. L., Lafont, F., Ríos, G., and Guilbert, S. 1989. Osmotic dehydration. Study of mass transfer in terms of engineering properties. In *Drying '89,* edited by A. S. Mujumdar and M. Roques. Hemisphere Publishing, New York.

Schwartzberg, H. 1982. *Freeze Drying*—Lecture notes. Food Engineering Department, University of Massachussets. Amherst, MA.

USDA. 1994. *Agricultural statistics 1994.* National Agricultural Statistics Service. Washington, D.C.

Van Arsdel, W. B. and Copley M. J. 1963. *Food Dehydration, Vol 1. Principles.* AVI Publishing, Westport, CT.

CHAPTER 2

FUNDAMENTALS OF AIR-WATER MIXTURES AND IDEAL DRYERS

2.0 INTRODUCTION

Air–water mixtures are used in most drying operations. The thermodynamic properties of this type of mixture are reported elsewhere (Himmelblau, 1982), but we consider it important to include in this chapter fundamental aspects of an air–water mixture before entering into the details of drying operations such as spray drying, fluidized bed drying, or atmospheric drying.

Some of the topics discussed in this chapter include moisture content, dry bulb temperature, relative humidity, humid heat, dew point, saturation condition, adiabatic saturation, and wet bulb temperature, with examples to clarify the concepts. In addition, a brief introduction to ideal dryers is presented.

2.1 FUNDAMENTALS OF AIR–WATER MIXTURES

The term *humidification* has been used in engineering to describe the interphase transfer of mass and energy between a gas and a pure liquid when they are brought into contact. The term covers not only humidification but also dehumidification of the gases, measurement of vapor content, and cooling of both gases and liquids.

For air–water mixtures the term *vapor* is related to water whereas the term *gas* is related to air. Air is considered, for all purposes, as a single substance although it is really a mixture of gases such as nitrogen, oxygen, argon, carbon dioxide, neon, helium, and other minor components. Its physical and chemical properties are reported elsewhere (Geankoplis, 1983).

2.1.1 Ideal Gas Relationships

The ideal gas law is used to predict the behavior of air–water mixtures because air temperature is high enough, and water vapor pressure is low enough in relation to their respective saturation points. Thus, their pressure–volume–temperature (PVT) relationship can be expressed as (Himmelblau, 1982):

$$P_{air}V = n_{air}RT \tag{1}$$

$$P_{water}V = n_{water}RT \tag{2}$$

where P_{air} is the partial pressure of air in the mixture, P_{water} is the partial pressure of water vapor, V is the total volume, R is the gas constant, T is the absolute temperature, n_{air} is the number of moles of air, and n_{water} is the number of moles of water. The total pressure of the air–water vapor mixture can be expressed as:

$$P_{total} = P_{air} + P_{water} \tag{3}$$

which is known as Dalton's Law. Dalton postulated that the sum of the partial pressures exerted by each component is equal to the total pressure of the system. A similar relationship may be used for volumes when the system is at a constant temperature and pressure.

EXAMPLE 1

An air–water mixture is confined in a vessel at 50°C and 101.325 kPa. The volume of the vessel is 10 m³ with 0.078 kg-mol of water. Determine the total number of moles in the vessel, the number of moles of air, and the partial pressure of water and air in the vessel.

Answer

The total number of moles in the vessel is evaluated from the ideal gas law for the mixture as follows:

$$P_{mix}V = n_{mix}RT$$

Replacing:
> $P_{mix} = 101.325$ kPa
> $V = 10$ m³
> $T = 323$ K
> $R = 8315$ N m/(kg-mol K)

Rearranging the ideal gas law:

$$n_{mix} = P_{mix}V/(R\,T)$$

$$n_{mix} = 0.377 \text{ kg-mol}$$

The moles in the mixture can be separated into moles of air (n_a) and water (n_w); then the moles of air can be calculated from the previous result:

$$n_a = n_{mix} - n_w$$

$$n_a = 0.378 - 0.078$$

$$n_a = 0.300 \text{ kg-mol of air}$$

The partial pressure of water is estimated using the ideal gas law:

$$P_w = n_wRT/V$$

$$P_w = (0.078 \text{ kg-mol})(8315 \text{ N m/kg-mol K})(323 \text{ K})/(10 \text{ m}^3)$$

$$P_w = 20{,}948 \text{ N/m}^2 \text{ or } 20.9 \text{ kPa}$$

The partial pressure of air can be determined by using Dalton's Law:

$$P_a = P_{mix} - P_w$$

$$P_a = 101{,}325 - 20{,}948$$

$$P_a = 80{,}377 \text{ N/m}^2 \text{ or } 80.4 \text{ kPa}$$

2.1.2 Moisture Content of Air

In drying operations, the properties of moist air change as a function of time. The main change is the amount of water removed from the product while air passes across the system. It is convenient to express the change in moist air properties in terms of dry air. Equations (1) and (2) can be combined to express the amount of water in terms of the amount of dry air, which results in the following expression for the molal absolute humidity:

$$W' = \frac{n_{water}}{n_{air}} = \frac{P_{water}}{P_{air}} \tag{4}$$

where W' is expressed in moles of water per mole of dry air. The humidity ratio or absolute humidity is obtained when W' is expressed as the water/air mass ratio rather than moles:

$$W = W'\left(\frac{M_{water}}{M_{air}}\right) \tag{5}$$

where M_{water} is the molecular weight of water and M_{air} is the molecular weight of air. Equation (4) is preferred over Eq. (5) because the moles and volumes can be interrelated easily through the ideal gas law.

The mole fraction of the water in an air-water mixture is defined as:

$$y_{water} = \frac{n_{water}}{\left(n_{air} + n_{water}\right)} \tag{6}$$

or

$$y_{water} = \frac{P_{water}}{\left(P_{air} + P_{water}\right)} \tag{7}$$

Notice the numerators of Eqs. (6) and (7) can be obtained by adding Eqs. (1) and (2) or by Dalton's Law:

$$(P_{air} + P_{water})V = (n_{air} + n_{water})RT \tag{8}$$

and Eq. (8) can be summarized as:

$$P_{total}V = n_{total}RT \tag{9}$$

If the mole fraction of water is known and the total pressure is known, it is possible to evaluate the partial pressure of water:

$$P_{water} = y_{water}P_{total} \tag{10}$$

and a similar expression can be written for the air:

$$P_{air} = y_{air}P_{total} \tag{11}$$

2.1.3 Psychrometric Chart

The psychrometric chart (Figure 2.1) for an air–water mixture is widely used because it relates basic properties such as humid volume, enthalpy, and humidity. It is necessary to understand the following terms in order to interpret the chart:

2.1.3.1 Dry Bulb Temperature

This is the temperature of the mixture measured by the immersion of a thermometer in the mixture without any modification on the thermometer.

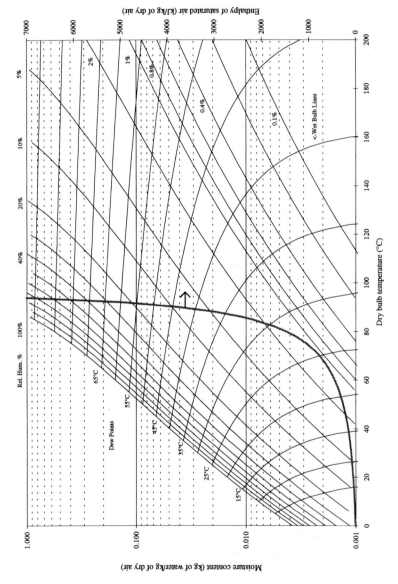

Figure 2.1. Psychometric chart for an air–water mixture.

14

2.1.3.2 Relative Saturation or Relative Humidity

The relative humidity is defined as the ratio between the partial pressure of water vapor (P_{water}) in the system and the partial pressure of water vapor ($P_{water\text{-}sat}$) in a saturated condition at the same temperature of the system. It can be expressed as:

$$\Phi = 100\,\frac{P_{water}}{P_{water\text{-}sat}} = 100\,\frac{X_{water}}{X_{water\text{-}sat}} \tag{12}$$

where X_{water} is the mole fraction of water in the mixture and $X_{water\text{-}sat}$ is the mole fraction of water in a saturated mixture at the same temperature. The saturation line for $P_{water\text{-}sat}$ in the psychrometric chart is identified as 100% relative humidity while other humidity levels are identified with their respective percentages.

2.1.3.3 Percentage Saturation or Percentage Absolute Humidity

The percentage saturation is expressed as:

$$\mu = 100\,\frac{W'}{W'_{sat}} \tag{13}$$

where W'_{sat} is the molal absolute humidity as defined by Eq. (4) for saturated conditions at the dry bulb temperature of the mixture.

2.1.3.4 Humid Volume or Specific Volume (ω)

Humid volume or specific volume is defined as the volume of unit mass of dry gas and the accompanying vapor at the mixture temperature and pressure, in m3/kg of dry air. It can be expressed for an air–water mixture as (Treybal, 1980):

$$v_h = T(2.83 \times 10^{-3} + 4.56 \times 10^{-3}W) \tag{14}$$

where T is the absolute temperature of the mixture (K), and W is the absolute humidity.

2.1.3.5. Humid Heat

The term *humid heat* represents the amount of heat required to raise the temperature of 1 kg of dry air plus the water vapor present by 1°C (Geankoplis, 1983). It can be defined for an air–water vapor mixture as:

$$C_s = C_{air} + WC_{water} \tag{15}$$

or

$$C_s = 1.005 + W1.884 \left(\frac{kJ}{kg \text{ of dry air K}} \right) \tag{16}$$

where W is the absolute humidity, C_{air} is the heat capacity of air (kJ/kg of dry air – K) and C_{water} is the heat capacity of water (kJ/kg of water – K).

2.1.3.6. Enthalpy of a Vapor–Gas Mixture

The enthalpy of a vapor–gas mixture is defined as the sum of the enthalpies of the gas and vapor content. This relationship can be expressed for air–water mixtures as:

$$H' = C_s(T - T_0) + \lambda_0 W \tag{17}$$

where T_0 is a reference temperature and λ_0 is the latent heat of water at T_0. The value of λ_0 is 2501.4 kJ/kg of water using air and saturated water vapor at 0°C as the reference point.

2.1.3.7 Dew Point

The dew point of an air–water mixture is the temperature at which the mixture becomes saturated when cooled at constant total pressure. If the mixture is cooled to a temperature below the dew point, the mixture will condense water. The dew point temperature can be determined from the psychrometric chart by drawing a straight line from a given point until reaching the saturation line, and the corresponding dry bulb temperature is the dew point.

2.1.4 The Saturated Condition

When a gas holds the maximum amount of vapor it is said to be in a saturated state. The partial pressure of the water vapor at saturated conditions can be found in a standard steam table or by using mathematical models such as (Treybal, 1980):

$$\ln(P_{\text{water-sat}}) = (-5674.53359/T) + 6.3925 - 0.9678 \times 10^{-2}T$$
$$+ 0.6222 \times 10^{-6}T^2 + 0.2075 \times 10^{-8}T^3$$
$$+ 0.9484 \times 10^{-12}T^4 + 4.1635 \ln(T) \qquad (18)$$

for a temperature range of −100 to 0°C, with T in K, and $P_{\text{water-sat}}$ in Pascals (Pa):

$$\ln(P_{\text{water-sat}}) = (-5800.2206/T) + 1.3915 - 0.0486T$$
$$+ 0.4176 \times 10^{-4}T^2 - 0.1445$$
$$+ 10^{-7}T^3 + 6.546 \ln(T) \qquad (19)$$

for a temperature range of 0 to 200°C.

The moisture content at saturation, W_{sat}, can be described by (Treybal, 1980):

$$W_{\text{sat}} = 0.621 \frac{P_{\text{water-sat}}}{\left(1.0133 \times 10^5 - P_{\text{water-sat}}\right)} \qquad (20)$$

Notice that Eq. (20) also applies to non-saturated conditions ranging from zero to saturation. For the nonsaturated state, the term $P_{\text{water-sat}}$ is replaced by the partial pressure of water (P_{water}). Equation (20) can be expressed as a function of the mole fraction of water and air (Treybal, 1980):

$$W_{\text{sat}} = 0.621 \frac{X_{\text{water}}}{X_{\text{air}}} \qquad (21)$$

2.1.5 Adiabatic Saturation Temperature

Liquid water at temperature T_{sat} recirculated through an adiabatic chamber is said to be at adiabatic saturation temperature when an entering gas at temperature T and humidity W is saturated at temperature T_{sat} and humidity W_{sat}. The enthalpy balance over the chamber (Figure 2.2) can be expressed as follows:

$$C_s(T - T_0) + W\lambda_0 = C_s(T_{sat} - T_0) + W_{sat}\lambda_0 \qquad (22)$$

where C_s is the humid heat, T_0 is the reference temperature of 0°C, and λ_0 is the latent heat of water (2501.4 kJ/kg of water). Equation (22) can be rearranged by using Eq. (16) as follows:

$$\frac{\left(W - W_{sat}\right)}{\left(T_{sat} - T\right)} = \frac{\left(1.005 + 1.884W\right)}{2501.4} \qquad (23)$$

2.1.6 Wet Bulb Temperature

The adiabatic saturation temperature is attained when large amounts of water come in contact with the entering gas. When a small amount of water is exposed to a con-

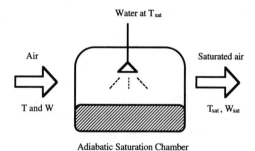

Adiabatic Saturation Chamber

Figure 2.2.   Adiabatic saturation chamber. (Adapted from Geankoplis, 1983.)

tinuous stream of gas under an adiabatic condition, it reaches a steady-state nonequilibrium temperature known as the wet bulb temperature.

The thermodynamic wet bulb temperature can be formally defined as the temperature T_{wb} at which water, by evaporating into moist air at a given dry bulb temperature T and moisture content W, can bring air to saturation adiabatically while constant pressure is maintained. The latent heat required for the evaporation will be supplied at the expense of the sensible heat of the liquid, and the temperature of the liquid decreases. The heat and mass balance can be used to describe the wet bulb temperature as follows:

$$q_t = q_s + N_{water}\lambda_{water} + N_{air}\lambda_{air} \qquad (24)$$

where q_t is the total heat transfer, N_{air} is the air mass flux, q_s is the sensible heat transfer flux, and N_{water} is the water vapor mass flux. The quantities q_s and N_{water} can be expressed as:

$$q_s = h_{air}(T_{air} - T_{wb}) \qquad (25)$$

$$N_{water} = k_{air}(P_{water\text{-}T} - P_{water\text{-}wb}) \qquad (26)$$

while q_t and N_{air} are zero. Incorporating Eqs. (25) and (26) into Eq. (24):

$$T_{wb} = T_{air} - \frac{\lambda_{water}\left(W_{wb} - W_T\right)\left(M_{air}P_{air}k_{air}\right)}{h_{air}} \qquad (27)$$

where h_{air} and k_{air} are the convective heat transfer and gas-phase mass transfer coefficients of the air. The ratio $h_{air}/M_{air}P_{air}k_{air}$ is reported to be 950N m/kg K for the air–water vapor system (Treybal, 1980).

The wet bulb temperature is similar to the adiabatic saturation temperature but with replacement of C_s by the ratio $h_{air}/M_{air}P_{air}k_{air}$. For many practical purposes the adiabatic saturation curves of the psychrometric chart can be used instead of Eq. (27).

Wet bulb temperature is determined with a thermometer whose bulb has been covered with a wet cloth. The thermometer is immersed in a rapidly moving stream of air, and the temperature will reach a value lower than the dry bulb temperature of the air if the latter is unsaturated. Figure 2.3 illustrates the method to measure T_{wb} and Figure 2.4 shows an example of a psychrometer. A psychrometer is usually used to evaluate both the dry bulb and wet bulb temperatures. It consists of two thermometers, one of which has a wick or cloth thoroughly wet with water. The psychrometer is exposed to the air–water vapor stream, the temperatures of the mixture are measured, and from a knowledge of these values the humidity is computed.

The state of a given air–water mixture is commonly specified by the wet bulb temperature and dry bulb temperature. Properties such as dew point, humidity ratio, relative humidity, and enthalpy can be evaluated using the dry bulb and wet bulb temperatures from the psychrometric chart as previously explained.

The heat transfer to or from air may result in a change in the dry bulb temperature of the mixture. A change in the humidity conditions may also occur as a result of latent heat added or removed from the mixture.

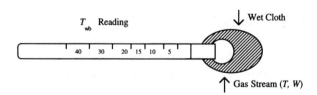

Figure 2.3.   Wet bulb temperature measurement. (Adapted from Himmelblau, 1982.)

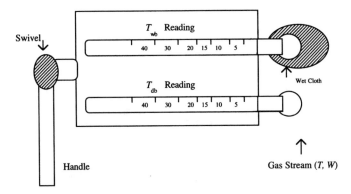

Figure 2.4.   Wet bulb–dry bulb psychrometer. (Adapted from White and Roos, 1991.)

The quantity of heat added or removed may be obtained by the difference in enthalpy between the initial and final conditions, while the amount of water added or removed may be calculated by differences between the humidity ratios before and after the process. Note that when the saturation line is reached the air is not able to retain or absorb more water, and the process moves along the saturation line. The case of cooling air below its dew point is an example of water leaving the mixture as a condensate while the saturation condition is attained. The concepts dew point, dry bulb temperature, wet bulb temperature, and adiabatic saturation line are summarized in Figure 2.5.

EXAMPLE 2

Determine the absolute humidity, T_{db} and T_{wb}, of moist air with a dew point of 40°C and a relative humidity of 50%.

Solution

From the psychrometric chart, the absolute humidity is obtained by drawing a line from 40°C (dew point temper-

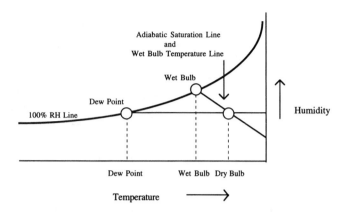

Figure 2.5.   Representation of moist air properties in the psychrometric chart.

ature) that intercepts the 100% relative humidity line, then reading the humidity as 0.049 kg of H_2O/kg of dry air. The T_{db} is obtained by extending the line at 0.049 kg of H_2O/kg of dry air until it intercepts the curve for 50% relative humidity, and then reading the temperature which, in this case, is 53°C. Finally, the T_{wb} is obtained by drawing a line that parallels the wet bulb lines until it intercepts the 100% relative humidity curve, and reading the wet bulb temperature which is 42°C. Figure 2.6 summarizes the above explanation.

2.2 MIXING TWO STREAMS OF AIR

The use of recycled air during a drying process results in a less expensive operation. In most cases, a hot/wet air stream is partially reused and mixed with fresh air. The mass and energy balance can be expressed as follows:

$$F_f W_f + F_h W_h = (F_f + F_h)W_m \tag{28}$$

$$F_f H_f' + F_h H_h' = (F_f + F_h)H_m' \tag{29}$$

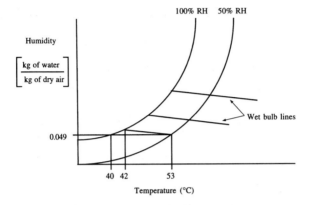

Figure 2.6. Psychrometric chart determinations for T_{wb}, T_{db}, and humidity.

where F is the air flow rate, W is the humidity, H' is the enthalpy, f represents fresh, h represents hot, and m represents the mixture of both fresh and hot. The above can be illustrated using the psychrometric chart as presented in Figure 2.7. The enthalpy differences between the mixture and each of the sources of air are proportional to the flow rates (Cook and DuMont, 1991).

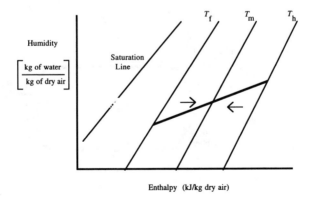

Figure 2.7. Representation of the mixing of two air streams.

2.3 HEAT AND MASS BALANCES IN IDEAL DRYERS

2.3.1 Continuous Dryer Without Recirculation

A continuous-type dryer, where the drying air flows countercurrently to the solids flow, is presented in Figure 2.8. The solids enter at a rate of F_s (kg of dry solid/h), having a moisture content of w_{s1} and a temperature T_{s1}. The solids leave the dryer at w_{s2} and T_{s2}. Air enters at a rate of F_a (kg of dry air/h) with a temperature of T_{a1} and having a humidity of W_{a1} (kg of water/kg of dry air). It leaves at T_{a2} and W_{a2}.

The material balance on the moisture is as follows:

$$F_a W_{a1} + F_s w_{s1} = F_a W_{a2} + F_s w_{s2} \tag{30}$$

and the heat balance on the dryer is:

$$Q_1 + F_a H'_{a1} + F_s H'_{s1} = Q_L + F_a H'_{a2} + F_s H'_{s2} \tag{31}$$

where Q_1 is the heat added to the dryer from any other external source, Q_L is the heat loss, and H'_m is the enthalpy of the air as expressed in Eq. (15):

$$H'_a = C_s(T_a - T_0) + W\lambda_0 \tag{32}$$

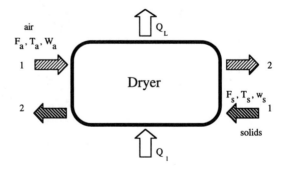

Figure 2.8.   Ideal continuous dryer without recirculation.

where C_s is given in Eq. (17). The enthalpy of the solid is given by:

$$H'_s = C_{ps}(T_s - T_0) + WC_{pa}(T_s - T_0) \tag{33}$$

where C_{ps} is the heat capacity of the solids, and C_{pa} is the heat capacity of liquid moisture (4.187 kJ/kg K).

EXAMPLE 3

Consider the dryer in Figure 2.8 being used to dry 500 kg of solids/h (dry solids) containing 0.06 kg of water/kg of dry solids to a value of 0.005 kg of water/kg of dry solids. The solids enter at 30°C and leave at 70°C, with the heat capacity (C_{ps}) being 1.465 kJ/kg K. The air enters at 100°C with a humidity of 0.010 kg of water/kg of dry air and leaves at 45°C. Calculate the air flow and outlet humidity, assuming $Q_1 = QL = 0$.

Answer

Using Eq. (30) for the mass balance:

$$F_a(0.01) + 500 * 0.06 = F_a W_{a2} + 500 * 0.005$$

or $\qquad F_a = -27.5/(0.01 - W_{a2})$

Using Eqs. (31), (32), and (33) for the heat balance:

Air:

$$H'_a = C_s(T_a - T_0) + W\lambda_0$$

$$H'_{a1} = [1.005 + 1.88 \cdot 0.01] \cdot (100 - 0) + (0.01 \cdot 2501)$$
$$= 127.39 \text{ kJ/kg of dry air}$$

$$H'_{a2} = [1.005 + 1.88 \cdot W_{a2})] \cdot (45 - 0) + (W_{a2} \cdot 2501)$$
$$= 45.23 + 2585.6 \cdot W_{a2} \text{ kJ/kg of dry air}$$

Solids:

$$H'_s = C_{ps}(T_s - T_0) + WC_{pa}(T_s - T_0)$$

$$H'_{s1} = 1.465 \cdot (30 - 0) + 0.06 \cdot 4.187 \cdot (30 - 0)$$
$$= 51.48 \text{ kJ/kg of dry solid}$$

$$H'_{s2} = 1.465 \cdot (70 - 0) + 0.005 \cdot 4.187 \cdot (70 - 0)$$
$$= 104.02 \text{ kJ/kg of dry solid}$$

Replacing the enthalpy values in Eq. (31):

$$F_a \cdot 127.39 + 500 \cdot 51.49$$
$$= F_a \cdot (45.23 + 2585.6 \cdot W_{a2}) + 500 \cdot 104.02$$

Solving the mass and heat balance simultaneously:

$$F_a = 1833.33 \text{ kg of dry air/h}$$

$$H_{a2} = 0.025 \text{ kg of water/kg of dry air}$$

2.3.2 Continuous Dryer With Recirculation

The recirculation of the drying air is used to reduce costs and control humidity. Hot air leaving the dryer is recirculated and combined with fresh air as shown in Figure 2.9. A material balance on the heater leads to:

$$F_{a1}W_{a1} + F_{a6}W_{a2} = (F_{a1} + F_{a6})W_{a4} \qquad (34)$$

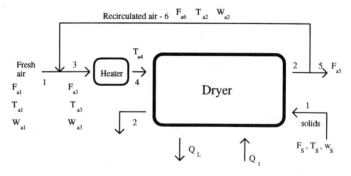

Figure 2.9.   Ideal continuous dryer with recirculation.

where F_{a1} is the fresh air rate, F_{a6} is the recirculation air rate, W_{a1} is the moisture of the fresh air, and W_{a2} is the moisture content of the air leaving the dryer. A similar balance can be done at the dryer:

$$(F_{a1} + F_{a6})W_{a4} + F_s w_{s1} = (F_{a1} + F_{a6})W_{a2} + F_s w_{s2} \quad (35)$$

The heat balances can be made for the heater, dryer, and overall system.

2.4 CONCLUDING REMARKS

The fundamentals of air–water vapor mixture properties (dry bulb temperature, wet bulb temperature, enthalpy, volume, adiabatic saturation, etc.) have been presented. The use of a psychrometric chart and the meaning of saturated state, humidity ratio, and relative humidity now should be familiar to the reader. The basic concepts of heat and mass transfer balances for ideal dryers have also been presented. They are discussed in more detail in the following chapters, where specific applications such as spray drying, freeze drying, and atmospheric drying are covered.

2.5 REFERENCES

Cook, E. M. and DuMont, H. D. 1991. *Process Drying Practice*. McGraw-Hill, New York.

Geankoplis, C. J. 1983. *Transport Processes and Unit Operations*, Second edition. Allyn and Bacon, Newton, MA.

Himmelblau, D. M. 1982. *Basic Principles and Calculations in Chemical Engineering*, Fourth edition. Prentice-Hall International Series. Prentice-Hall, Englewood Cliffs, NJ.

Treybal, R. E. 1980. *Mass-Transfer Operations*, Third edition. McGraw-Hill Chemical Engineering Series, edited by J. V. Brown and M. Eichberg. McGraw-Hill, New York.

White, G. M. and Roos, I. J. 1991. Humidity. In *Instrumentation and Measurement for Environmental Sciences*. Third edition, edited by Z. A. Henry, G. C. Zoerb, and G. S. Birth. ASAE, St. Joseph, MI.

CHAPTER 3

PHYSICAL, CHEMICAL, AND MICROBIOLOGICAL CHARACTERISTICS OF DEHYDRATED FOODS

3.0 INTRODUCTION

The preservation of the initial cellular structure of the material being dried is, in most cases, a function of the process applied to the food. This consideration applies not only to drying but also to blanching and freezing. In the removal of water from high-moisture containing products such as fruits and vegetables, it is important to remember that cell membranes may or may not been damaged during processing (Le Maguer, 1987).

Drying of food products not only affects the water content of the product, but also alters other physical, biological, and chemical properties such as enzymatic activity, microbial spoilage, crispiness, viscosity, hardness, aroma, flavor, and palatability of the foods. This chapter deals mainly with these properties and the effect drying has on them. Special attention is given to water activity, sorption isotherms, glass transitions, and food structure, given their role in food quality, safety, and stability.

3.1 WATER CONTENT OF FOODS

Free or unbound water is defined as water within a food that behaves as pure water. Unbound water is removed during the constant rate period of drying when the nature of the food does not have a great effect on the drying process.

Okos et al. (1992) and Leung (1986) defined the term *bound water* as water that exhibits a lower vapor pressure, lower mobility, and greatly reduced freezing point than pure water. Bound water molecules have different kinetic and thermodynamic properties than ordinary water molecules. Bound water can be determined by measuring water that cannot be frozen at subfreezing temperatures with nuclear magnetic resonance (NMR) or differential thermal analysis (DTA). Also, the use of dielectric properties of water to determine the bound water content is reported in the literature. Dielectric properties of water depend on the mobility of the molecules in response to changes in the applied electric field. The properties of bound water range between the rigidly held dipoles of ice and the much more mobile liquid water molecules (Karel, 1975a; Troller and Christian, 1978; Leung, 1986). The relationship between the water and chemical species contained in foods, which are reflected in the sorption properties of foods, is summarized in terms of the types of interactions between them as follows (Van den Berg, 1985):

- London–van der Waals dispersion forces
- Hydrogen bonds
- Coulomb forces between water, ions, and dissociated groups
- Steric effects
- Dissolution effects
- Changes in mobility of polymer segments
- Capillary forces

and the water binding characteristics are defined as:

Tightly bound water   Water activity < 0.3
Moderately bound water   Water activity 0.3 to 0.7
Loosely bound water   Water activity > 0.7
Free water   Water activity ≈ 1.0

An example of the above interactions is observed when proteins are added to a food. Baked and meat products take advantage of some of the biophysical properties of proteins to improve the cohesion, water absorption, foaming characteristics, and browning of the formulation (Cheftel et al., 1985). It is reported that the tertiary structure of proteins is responsible for their water binding properties. Hydrophobic amino acids are located toward the inside of the protein molecule, and polar amino acids are found mainly at the surface in water-soluble proteins. The protein surface has sites where hydrogen bonds are established and that allow interactions with other molecules, especially water. This contributes to the specific structure and solubility of some proteins. Water absorbed into the protein represents up to 0.3 g/g of dry protein, and water that occupies the first two or three layer sites on the protein surface can represent an additional 0.3 g/g (Cheftel et al., 1985). The solubility of proteins depends mainly on pH, ionic strength, and temperature.

An objective of using fat in food formulation is to improve the binding properties of the mixture. Liquid crystals, or mesomorphic phases, formed by heating lipids in the solid state in the presence of water, are said to be lyotropics (Nawar, 1985). The water penetrates the ordered polar groups of the lipids and forms a mesomorphic phase or structure. The capacity of the structure to retain water depends on the composition of the lipids.

3.2 DETERMINATION OF WATER CONTENT

Total moisture content can be determined by a gravimetric method. Assuming only water leaves the sample, the percentage of moisture can be evaluated from the drying of a preweighed sample to a constant weight. In the case of food, special care is necessary to prevent case hardening (Troller and Christian, 1978) or decomposition of the sample so that results are not affected. Vacuum ovens are used in most moisture content determinations. Gas chromatography is also used to determine the total content of water in a sample. In general, the method consists of blending the sample in the presence of methanol and determining the peak area from the chromatogram. The value is then compared against a series of standards containing known amounts of water. Karl Fisher introduced a specific reagent for water that contained pyridine, methanol, sulfur dioxide, and iodine. The method is reported useful for the analysis of dried fruits and vegetables, candies, coffee, and fats (Troller and Christian, 1978). The method can be used on virtually all solid or semisolid agricultural and other food products. Water is extracted with anhydrous methanol mixed with a finely ground sample. After extraction, the solids are removed by filtration and the liquid is titrated with the Fisher reagent containing iodine, sulfur dioxide, and pyridine.

3.3 WATER ACTIVITY

A fundamental aspect considered in food preservation is how the water is bound into the food rather than the amount of water already in the food. The term *water activity* (a_w) was introduced in the 1950s to describe the state of water in food products (Monsalve-González et al., 1993a). Water activity is a key factor in microbial growth, toxin production, and enzymatic and nonenzymatic reactions (Leung, 1986). Most of the bacteria cannot grow below an a_w of 0.90 (Bone, 1987), the factor used as a regulatory parameter in the United States by the Food and Drug Administration pertaining to canned foods (Leung, 1986; FPI, 1982).

3.3.1 Thermodynamics of Water Activity

Water activity is a term used to indicate the relation between a food and the equilibrium relative humidity of the surrounding atmosphere (Van den Berg and Bruin, 1981). The Gibbs free energy of an ideal system is defined as (Leung, 1986):

$$G = E + PV - TS \tag{1}$$

where S is the entropy, E is the internal energy, T is the absolute temperature, P is the pressure, and V is the volume. Equation (1) can be expressed in a differential form for a multicomponent system as a function of T, P and n_i (moles of a particular component) as:

$$dG = V\, dP - S\, dT + \Sigma(\partial G/\partial n_i)_{T,P,njdn_i} \tag{2}$$

where $(\partial G/\partial n_i)_{T,P,n_j}$ is the partial molal Gibbs free energy or chemical potential (μ_i):

$$\mu_i = G_i = (\partial G/\partial n_i)_{T,P,njdn_i} \tag{3}$$

This implies that the chemical potential is the change in free energy of the system corresponding to an infinitesimal change in the number of moles in one component when other parameters are held constant. It indicates the escaping tendency of this particular component.

Equation (3) can be differentiated with respect to P and written as follows:

$$\left(\frac{\partial \mu}{\partial P}\right)_T = \frac{\partial}{\partial P}\left(\frac{\partial G}{\partial n_i}\right)_{T,P} = \frac{\partial}{\partial n_i}\left(\frac{\partial G}{\partial P}\right)_T = \left(\frac{\partial V}{\partial n_i}\right)_{T,P} = \bar{V}_i \tag{4}$$

where $\bar{V}_i$ is the partial molal volume considering the ideal gas equation. Combining Eq. (3) and (4), the chemical potential can be expressed as:

$$d\mu_i = RT\, d(\ln P_i) \tag{5}$$

or

$$\mu_i = \mu_i^0 + RT \ln\left(\frac{P_i}{P_i^0}\right) \tag{6}$$

Equation (5) can be written for real gases as:

$$d\mu_i = RT\, d(\ln f_i) \tag{7}$$

or after integration as:

$$\mu_i = \mu_i^0 + RT \ln\left(\frac{f_i}{f_i^o}\right) \tag{8}$$

μ_i^0 and f_i^0 are the chemical potential and fugacity of component i at the standard state. The activity coefficient can be expressed as:

$$a_i = \frac{f_i}{f_i^o} \tag{9}$$

Thus, Eq. (8) becomes:

$$\mu_i = \mu_i^0 + RT \ln a_i \tag{10}$$

and considering the limit of f_i/P when P tends to zero (Van Ness and Abbott, 1982):

$$\lim_{P \to 0}\left(\frac{f_i}{P}\right) = 1 \tag{11}$$

Eq. (9) can be expressed as:

$$a_i = \frac{P_i}{P_i^o} \tag{12}$$

valid for ideal gases or real gases at low pressure (Van Ness and Abbott, 1982).

A solution in equilibrium with its vapor phase has the same chemical potential in the two phases for each of the components:

$$\mu_i^l = \mu_i^g \tag{13}$$

Thus, the vapor–liquid equilibrium in a multicomponent system can be expressed as (Van Ness and Abbott, 1982):

$$f_i^v = \Phi_i^v y_i P \tag{14}$$

$$f_i^l = \Phi_i^l x_i P \tag{15}$$

where f_i^v and f_i^l are the fugacities of vapor and liquid phases, Φ_i^v and Φ_i^l are the fugacity coefficients for vapor and liquid phases, P is the system pressure, and y_i and x_i are the mole fractions of a particular component in the vapor and liquid phases, respectively. The liquid-phase activity coefficient is expressed as:

$$\gamma_i = f_i^l / x_i f_i^o \tag{16}$$

where f_i^o is a standard state fugacity of component i in the mixture. The activity coefficient accounts for deviations from the ideal solution behavior in the liquid phase. Equations (14) and (16) can be combined based upon vapor–liquid equilibrium (Van Ness and Abbott, 1982) as follows:

$$\gamma_i x_i f_i^o = \Phi_i^v y_i P \tag{17}$$

If it is assumed that the vapor phase behaves as an ideal gas and the fugacities of liquids are independent of pressure (Van Ness and Abbott, 1982), Eq. (17) can be expressed as:

$$y_i P = x_i P_i^{sat} \gamma_i \tag{18}$$

where γ_i is a function of the liquid phase composition and temperature but not of pressure.

3.3.1.1 Raoult's Law

The activity coefficient γ_i accounts for the nonideal behavior of the liquid phase. If it is assumed that the liquid phase behaves as an ideal solution the activity coefficient becomes unity and Eq. (18) is reduced to Raoult's Law:

$$x_i = y_i P / P_i^{sat} \qquad (19)$$

The ratio of partial pressure to saturation pressure, as expressed by Eq. (12), can be applied to Eq. (19) to reduce it to the following expression:

$$a_i = x_i \qquad (20)$$

Equation (20) is considered for dilute solutions (ideal mixtures).

Equations (12) and (20) can be expressed for water as:

$$a_w = P_w / P_w^o \qquad (21)$$

$$a_w = x_w \qquad (22)$$

where a_w is the water activity, P_w is the partial pressure of water in the vapor phase, P_w^o is the saturation pressure of pure water, and x_w is the molar fraction of water in the mixture.

The application of Raoult's Law to food systems is not practical because of solvent–solute interactions which promote significant deviations from an ideal solution (Stokes, 1979). Equation (22) can be expressed for nonideal solutions as:

$$a_w = \gamma_w x_w \qquad (23)$$

where γ_w is the activity coefficient of water which corrects for deviation from ideality.

3.3.2 Theoretical and Empirical Models to Predict a_w and Applications

Water activity can be estimated by different theoretical and empirical models considering the types of solutes in the solution (electrolytes, nonelectrolytes, or mixtures). The depression of water activity by dissolved solutes is one well-known factor in foods. Nonelectrolytes behave differently from electrolyte solutions because of the inter-action of charged molecules and the formation of ion clouds around each ion in the latter. Electrolytes deviate from ideal solutions even at very low concentrations (Leung, 1986).

3.3.2.1 Nonelectrolytic Solutions

Soluble solids, which do not dissociate when brought into solution, are termed nonelectrolytes. These types of substances, also called humectants, are used to bind water and allow food products to maintain a soft palat-able texture (Lindsay, 1985; Villar and Silvera, 1987). Some of the most relevant mathematical models used to estimate the water activity of nonelectrolytes are as fol-lows:

Money and Born Equation

The Money and Born (1951) empirical equation for calcu-lating water activity of sugar confections, such as jams, fondant creams, and boiled sweets, is expressed as:

$$a_w = 1.0/(1.0 + 0.27n) \qquad (24)$$

where n is the number of moles of sugar per 100 g of water. The equation is based on Raoult's Law but can be applied over a wide range of concentrations (Money and Born, 1951).

Grover Equation

The Grover model is another empirical approach to esti-mate water activity in candy formulations (Karel, 1975a). This method assigns a sucrose equivalent conversion fac-tor to each ingredient in a candy:

$$a_w = 1.04 - 0.1\Sigma_i s_i c_i + 0.0045\Sigma_i (s_i c_i)^2 \qquad (25)$$

where c_i is the concentration of component i and s_i is the sucrose equivalent for different ingredients such as lactose (1.0), invert sugar (1.3), 45DE corn syrup (0.8), and gelatin (1.3).

Norrish Equation

The equation proposed by Norrish (1966) takes into account a nonideal thermodynamic approach to determine the water activity of binary mixtures. The model is derived from Eq. (23), and the partial molar excess free energy of mixing is expressed as (Norrish, 1966):

$$\Delta F_w^E = \partial(N\Delta F_w^E)/\partial N_w \qquad (26)$$

and it is related to concentration by the following expression for a binary mixture:

$$\Delta F_w^E = C_{12}X_1X_2 \qquad (27)$$

where C_{12} is a constant and X_1 and X_2 are the mole fractions of the components in the mixture. The activity coefficient for water is then defined as:

$$\log \gamma_w = \Delta F_w^E \qquad (28)$$

or from Eqs. (26), (27), and (28):

$$\log \gamma_w = KX_2^2 \qquad (29)$$

where $K = 0.434C_{12}/RT$, R is the universal gas constant, and T is the absolute temperature. K values, listed in Table 3.1 for some sugars, polyols, amino acids, and amides, are obtained from the slope of the plot of $\log \gamma_w$ versus X_2^2. Finally, the water activity is estimated by substituting the expression for the activity coefficient into Eq. (23):

$$a_w = X_1\exp(-KX_2^2) \qquad (30)$$

This model is reduced to Raoult's Law if very dilute solutions are considered, or the K value is near zero (Chirife, 1987).

EXAMPLE 1

The water activity of a glucose–water solution (2.44:1 wt/wt) can be estimated by means of the Norrish equation. The mole fractions are: $X_1 = 0.804$ and $X_2 = 0.196$. The Norrish constant for glucose is 2.25. Substituting for the values in Eq. (29) results in an estimated water activity of 0.74. The experimental water activity of this mix-

Table 3.1.   Value of Norrish constant K for some sugars, polyols, amino acids, and amides.

Sugars	K
Sucrose	6.47 ± 0.06
Maltose	4.54 ± 0.02
Glucose	2.25 ± 0.02
Xylose	1.54 ± 0.04
Lactose	10.2
Polyols	
Sorbitol	1.65 ± 0.14
Glycerol	1.16 ± 0.01
Mannitol	0.91 ± 0.27
Propylene glycol	4.04
Arabitol	1.41
Amino acids and amides	
α-Amino-n-butyric acid	2.59 ± 0.14
β-Alanine	2.52 ± 0.37
Lactamide	-0.705 ± 0.066
Glycolamide	-0.743 ± 0.079
Urea	-2.02 ± 0.33
Glycine	-0.868 ± 0.11
Organic acids	
Citric acid	6.2
Malic acid	1.82
Tartaric acid	4.68

(From Chirife et al., 1980 and Chirife and Favetto, 1992).

ture is reported as 0.78 (Teng and Seow, 1981). The use of Raoult's Law gives a water activity estimate of 0.80. The deviation from an ideal solution is corrected by the activity coefficient as defined by Eq. (29).

3.3.2.2 Electrolytic Solutions

Electrolytes are substances that exist in the form of ions when they are in solution. The water activity of a solution is affected by all the ionic species that are released when an electrolyte is dissociated in a solution (Stokes, 1979). This is the main difference with respect to nonelectrolyte compounds.

The osmotic coefficient, Φ, is defined to simplify some thermodynamic calculations when working with the activity values and the activity coefficients of solvents. Both activity values and activity coefficients vary little from the ideal value of one, when considering a dilute solution of electrolytes (Stokes, 1979):

$$\Phi = -1000 \ln a_s / (M_s \Sigma v_r m_r) \qquad (31)$$

where M_s is the molecular weight of the solvent, m_r is the molality of each ion in solution, v_r is the number of moles of each ion in solution, and a_s is the activity of the solvent. The osmotic coefficient tends to zero at infinite dilution of all solutes (Stokes, 1979), which recognizes the fact that Raoult's Law (Eq. 21) becomes exact at an infinite dilution condition.

Pitzer Equation

Pitzer (1973) recognized the dependence of short-range forces in binary interactions on ionic strength, a dependence not considered in the Debye–Hückel model. The Debye–Hückel model is an approximation to obtain thermodynamic properties that is valid for the coulombic forces, but omits the direct effects of short range forces (Pitzer, 1979). The Debye–Hückel limiting law is exact only at an ionic strength of less than 0.01 M (Leung, 1986). The osmotic coefficient of a single electrolyte can be expressed, according to Pitzer (1973), as follows:

$$\Phi - 1 = \left| z_m z_x \right| F + 2m \left(\frac{v_m v_x}{v} \right) B_{mx} + 2m^2 \left[\frac{\left(v_m v_x \right)^{1.5}}{v} \right] C_{mx} \quad (32)$$

$$F = -A \frac{I^{0.5}}{\left(1 + bI^{0.5} \right)} \quad (33)$$

$$I = 0.5 \Sigma_i m_i z_i^2 \quad \text{and} \quad |z_m z_x| = \Sigma_i m_i z_i^2 / \Sigma_i m_i \quad (34)$$

$$B_{mx} = B_{mx}(0) + B_{mx}(1) \exp(-\alpha I^{0.5}) \quad (35)$$

where z_m and z_x are the charges of m and x ions, v_m and v_x are the respective number of ions, $v = v_m + v_x$, m is the solution molality, A is the Debye–Hückel coefficient with a value of 0.392 at 25°C for water, b is 1.2, α is 2.0, I is the ionic strength, B_{mx} is the second virial coefficient, $B_{mx}(0)$ and $B_{mx}(1)$ represent the short range binary interactions of a single electrolyte, and C_{mx} is the third virial coefficient (Pitzer, 1973; Pitzer and Mayorga, 1973). Table 3.2 shows Pitzer constants for selected electrolytes. This model has been successfully applied to free energy data of aqueous electrolytes with an ionic strength (I) up to 6 M (Pitzer, 1973).

Water activity can be evaluated by rearranging Eq. (31):

$$a_w = \exp(-0.01802 \Phi \Sigma_i v_i m_i) \quad (36)$$

EXAMPLE 2

The following example explains the use of the Pitzer model when an electrolyte 1:1 type is in solution (e.g., a solution of NaCl, 2.31 molal at 25°C):

Table 3.2.   Values for Pitzer and Bromley constants for some electrolytes.

Electrolytes	$B(0)$[a]	$B(1)$	C	B[b]
NaCl	0.0765	0.2664	0.00127	0.0574
LiCl	0.1494	0.3074	0.00359	0.1283
KCl	0.0483	0.2122	−0.0008	0.0240
HCl	0.1775	0.2945	0.00080	0.1433
KOH	0.1298	0.3200	0.0041	0.1131
KH_2PO_4	−0.0678	−0.1042	−0.1124	
NaOH	0.0864	0.2530	0.0044	0.0747
NaH_2PO_4	−0.0533	0.0396	0.00795	−0.0460

[a] $B(0)$, $B(1)$, and C are Pitzer constants.
[b] B is the Bromley constant.
(From Bromley, 1973; Pitzer and Mayorga, 1973)

The Pitzer's $B_{mx}(0)$, $B_{mx}(1)$, and C_{mx} parameters for NaCl are 0.0765, 0.2664, and 0.00127, respectively. The following can be defined for the ions: $z_{Na} = 1$, $z_{Cl} = 1$, $\nu_{Na} = 1$, and $\nu_{Cl} = 1$. The osmotic coefficient is evaluated from Eq. (31) as follows:

$$m_{Na} = 2.31 \quad m_{Cl} = 2.31 \quad I = 2.31 \quad |z_{Na}z_{Cl}| = 1.0$$

$$F = -0.211 \quad B_{NaCl} = 0.089 \quad \Phi = 1.001$$

Water activity is evaluated from Eq. (36):

$$a_w = \exp(-0.01802 \times 1.001 \times (2.31 + 2.31))$$

$$a_w = 0.92$$

The experimental water activity of this mixture is 0.92 (Teng and Seow, 1981).

Bromley Equation

Bromley's (1973) equation is based on the assumption that B_{mx} may be approximated as the sum of individual ion B values. The osmotic coefficient is defined as:

$$\Phi = 1 + 2.303[T_1 + (0.06 + 0.6B)T_2 + 0.5BI] \qquad (37)$$

$$T_1 = A|z_m z_x| \frac{\left[1 + 2\left(1 + I^{0.5}\right)\ln\left(1 + I^{0.5}\right) - \left(1 + I^{0.5}\right)^2\right]}{I\left(1 + I^{0.5}\right)} \qquad (38)$$

$$T_2 = \frac{\left(1 + 2aI\right)}{a\left(1 + aI\right)^2} - \frac{\ln\left(1 + aI\right)}{a^2 I} \qquad (39)$$

$$a = \frac{1.5}{|z_m z_x|} \qquad (40)$$

where T_1 and T_2 are used to simplify Eq. (37), A is equal to 0.511 at 25°C, and B values are as reported by Bromley (1973) for different salt solutions or can be estimated from the individual B values as (Bromley, 1973):

$$B = B_m + B_x + \delta_m \delta_x \qquad (41)$$

Table 3.2 shows some examples of the Bromley constants for selected electrolytes. Although less accurate than Pitzer's equation, Bromley's model is quite effective and could be useful for thermodynamic calculations of electrolyte solutions (Pitzer and Mayorga, 1973). The water activity of the NaCl solution previously discussed can be estimated by means of the Bromley model as follows:

EXAMPLE 3

The Bromley B parameter is equal to 0.0574 for the NaCl. The ionic strength (I) and the charge product $|z_m z_x|$ are the same as above. Substituting these values in Eqs. (38), (39), and (40) gives:

$$T_1 = -0.0615 \quad T_2 = -0.0226 \quad a = 1.5$$

The osmotic coefficient is evaluated from Eq. (37) and the water activity from Eq. (38):

$$\Phi = 1.0062 \quad a_w = 0.919$$

3.3.2.3 Multicomponent Mixtures

The addition of different water activity depressors is necessary to maintain a soft palatable texture of foods (Lindsay, 1985). Different levels of water activity can be obtained by mixing different types of solutes. The prediction of water activity in multicomponent mixtures has been reviewed by different authors such as Teng and Seow (1981), Norrish (1966), Pitzer and Kim (1974), Ferro-Fontán et al. (1980), Ferro-Fontán et al. (1981), and Ferro-Fontán and Chirife (1981). In most cases, the effects of specific solute–solute interactions are neglected and individual binary water activity data are used to predict the water activity of the mixture (Teng and Seow, 1981).

Ross Equation

The Ross equation has been tested and appears to provide a reasonable means of estimating water activity of multicomponent aqueous mixtures over the intermediate and high water activity ranges (Teng and Seow, 1981). The model is based upon the Gibbs–Duhem relationship at constant temperature and pressure:

$$\Sigma_i x_i d(\ln \gamma_i) = 0 \qquad (42)$$

where x_i is the mole fraction of a component in the mixture and γ_i is the activity coefficient of the component. Equation (42) can be expressed for water in a binary mixture as:

$$d(\ln a_{\text{water}}) = -(1/55.5)\Sigma_i \delta(\ln m_i \gamma_i^0) \qquad (43)$$

where m_i is the molality and γ_i^0 is the activity coefficient of the solute in a binary mixture. This expression is valid when the temperature and the pressure remain constant.

The Ross equation can be derived from Eq. (43) and expressed as:

$$a_{\text{w}} = \Pi_i (a_{\text{w}}^0)_i \qquad (44)$$

where a_w^o is the water activity of each component in a binary mixture at the same concentration and temperature as in the multicomponent system. Chirife et al. (1980) reported the possibility of using the Ross equation to determine the water activity of food mixtures of solutes such as salts, sugars, and polyols. Nevertheless, it has been shown that the model may give significant deviations at reduced water activities in solutions of strong electrolytes.

Ferro-Fontán–Benmergui–Chirife Equation

Ferro-Fontán, Benmergui, and Chirife (1980) reported a refinement of Eq. (44) for electrolytes expressed as:

$$a_w = \Pi_i \left(a_{wi}(I) \right)^{\frac{I_i}{I}} \tag{45}$$

$$I_i = 0.5 v_i m_i |z_m z_x| \tag{46}$$

where $a_{wi}(I)$ is the water activity of a binary solution of i at the same total ionic strength (I) of the multicomponent solution, I_i is the ionic strength of component i in the mixture, v_i is the total of ions of the solute i in solution, m_i is the molality of the solute, and z_m and z_x are the charges of m and x ions.

EXAMPLE 4

The following example, using data reported by Tsong and Seow (1981) illustrates how to use the Ross and Ferro-Benmergui-Chirife equations.

An aqueous solution of NaCl (9.29% wt/wt) and LiCl (1.57% wt/wt) has an experimental water activity of 0.9312. The individual water activities can be obtained from experimental measurements of solutions at the same ionic strength as the mixture. In this example, the total ionic strength is 2.19, the NaCl ionic strength is 1.78, and the LiCl ionic strength is 0.41 (as evaluated by Eq. 46). The estimated water activities, by means of the Pitzer and Bromley equations for individual compounds, are as follows:

1. Water activity estimate using the Ross equation:

NaCl: $I_i = 1.78$ a_w(Pitzer) = 0.939; a_w(Bromley) = 0.942

LiCl: $I_i = 0.41$ a_w(Pitzer) = 0.985; a_w(Bromley) = 0.984

Replacing the individual water activities in Eq. (44), the overall water activity of the mixture can be obtained:

a_w(ΠPitzer) = 0.926 a_w(ΠBromley) = 0.928

Notice that ΠPitzer and ΠBromley refer to the product of the individual water activities evaluated by the same model (Pitzer or Bromley).

2. Water activity estimate using the Ferro–Benmergui–Chirife model:

The prediction of water activities at the same ionic strength of the solution is based on the following expression:

$$m_i(I_T) = m_i \cdot (I_T/I_i) \qquad (47)$$

where I_T is the ionic strength of the mixture, $m_i(I_T)$ is the solute molality which affords an ionic strength equal to that of the mixture, I_i is the component i ionic strength in the mixture, and m_i is the actual molality of component i. The required constants and values to apply to the model are:

a. NaCl: $I_i = 1.78$; $I_T = 2.19$; $(I_i/I_T) = 0.81$
   a_w(Pitzer-I_T) = 0.923; a_w(Bromley-I_T) = 0.929

b. LiCl: $I_i = 0.41$; $I_T = 2.19$; $(I_i/I_T) = 0.18$
   a_w(Pitzer-I_T) = 0.901; a_w(Bromley-I_T) = 0.923

Incorporating the values in Eq. (45), the water activity of the mixture is obtained:

Ferro–Benmergui–Chirife (Bromley-I_T) = 0.928

Ferro–Benmergui–Chirife (Pitzer-I_T) = 0.920

where Bromley-I_T and Pitzer-I_T refer to the use of the individual water activities evaluated by each model.

Ferro–Chirife–Boquet Equation

Ferro-Fontán et al. (1981) reported a model developed based on Equations (29) and (45). Equation (29) is integrated into Eq. (45) for each component in the mixture, resulting in the following equation:

$$\left(a_w\right)_m = X_1 \exp\left[\left(-\frac{\sum\limits_{s=1}^{n} K_s m_s}{m}\right) X_2^2\right] \tag{48}$$

where K_s is the Norrish K value for each solute in the mixture, m_s is the molality of the solutes in the mixture, m is the total molality of the mixture, n is the number of species in the mixture, and X_1 and X_2 are the molar fractions of the solvent and solutes. This equation is similar to the Norrish (1966) equation but Ferro-Fontán et al.(1981) transformed the term K (Norrish equation) into:

$$\frac{\sum\limits_{s=1}^{n} K_s m_s}{m} \tag{49}$$

to relate the weight ratio and molecular weight of each component in the mixture. The final set of equations for nonelectrolyte solutions is:

$$(a_w)_m = X_1 \exp(-K_m X_2^2) \tag{50}$$

$$K_m = \sum_{s=1}^{n} K_s C_s \left(\frac{M_t}{M_s} \right) \tag{51}$$

$$M_t = \left[\sum_{s=1}^{n} \left(\frac{C_s}{M_s} \right) \right]^{-1} \tag{52}$$

where C_s is the weight ratio of each solute to the total of solids in the mixture, and M_s is the molecular weight of each component.

EXAMPLE 5

This example shows how to apply the Ferro–Chirife–Boquet model. A solution of glucose (5.96% wt/wt) and sucrose (46.01% wt/wt) has a measured water activity of 0.926 (Teng and Seow, 1981). The K values of glucose and sucrose are 2.25 and 6.47, respectively. The K_m and M_t values are evaluated from Eqs. (47) and (48).

$$K_m = 5.65 \quad M_t = 310.82$$

The predicted water activity is 0.923.

Lang–Steinberg Equation

Lang and Steinberg (1981) developed a model to describe the effect of nonsolutes (i.e., starch, casein, soy flour) on water activity. The model is based on the relationship between the moisture content and water activity of each component in the food. Lang and Steinberg (1981) considered the Smith equation to describe food sorption isotherm data. The Smith model is expressed as:

$$\ln(1 - a_w) = (X - a_i)/b_i \tag{53}$$

where a_i and b_i are evaluated from the plot of M versus $\ln(1 - a_w)$ and X is the moisture content. The overall water activity of the mixture can be estimated by the following equation:

$$\log(1 - a_w) = (MW - \Sigma a_i w_i)/\Sigma b_i w_i \tag{54}$$

where M is the moisture content of the mixture, W is the total dry material of the mixture, w_i is the dry material of each component i, and a_i and b_i are the Smith equation constants (intercept and slope) for each component i in the mixture. The model may predict water activities over the range 0.30 to 0.95 of a multicomponent mixture of known composition as discussed by Lang and Steinberg (1981).

EXAMPLE 6

The following example was discussed by Lang and Steinberg (1981). A mixture of starch (0.79 g/g of total sample) and sucrose (0.09 g/g of total sample) has a moisture content of 0.29 g of water/g of solids at an a_w of 0.90. The Smith constants for each component are (Lang and Steinberg, 1981):

$$\text{Starch: } a_i = 0.0989 \quad b_i = -0.1485$$

$$\text{Sucrose: } a_i = -0.5944 \quad b_i = 1.2573$$

Rearranging the data in terms of the Lang and Steinberg model:

$$M = 0.29 \text{ g of water/g of solid} \quad W = 0.89 \text{ g of solids}$$

$$(a_i w_i)_{\text{starch}} = 0.08 \text{ g of water} \quad (a_i w_i)_{\text{sucrose}} = -0.06 \text{ g of water}$$

$$(b_i w_i)_{\text{starch}} = -0.12 \text{ g of water} \quad (b_i w_i)_{\text{sucrose}} = 0.13 \text{ g of water}$$

Replacing the above values in Eq. (54) and taking the $\log_{10}$ to base 10:

$$a_w = 0.90$$

Pitzer–Kim Equation

Pitzer and Kim (1974), on the basis of the Debye–Hückel model, defined a working expression for the osmotic coefficient of mixed electrolytes:

$$\Phi - 1 = (\Sigma m_i)^{-1} \{ (IF' - F) + \Sigma_i \Sigma_j m_i m_j (\lambda_{ij} + I\lambda'_{ij}) \}$$

$$+ 2\ \Sigma_i\Sigma_j\Sigma_k m_i m_j m_k \mu_{ijk}\} \tag{55}$$

where $F' = dF/dI$, $\lambda'_{ij} = d\lambda_{ij}/dI$, m is the molality of a particular ion (i, j, or k), λ_{ij} is the second virial coefficient, and μ_{ijk} is the third virial coefficient. At very low concentrations of mixed electrolytes, the osmotic coefficient can be expressed as:

$$\Phi - 1 = -2\ AI^{1.5}/\Sigma_i m_i \tag{56}$$

Example 7 is presented to explain the use of the Pitzer and Kim model.

EXAMPLE 7

An aqueous solution of NaCl (9.29% wt/wt) and LiCl (1.57% wt/wt) has an experimental water activity of 0.9312 (Tsong and Seow, 1981). Equation (55) can be expressed as (Pitzer and Kim, 1974):

$$\Phi - 1 = (\Sigma m_i)^{-1}\{2IF + 2\ \Sigma_i\Sigma_j m_i m_j(B_{ij} + C_{ij}(\Sigma mz)/(z_i z_j)^{0.5}\} \tag{57}$$

where B_{ij} is evaluated by Eq. (35), F by Eq. (33), and I by Eq. (34):

$$I = 2.198 \qquad F = -0.2087$$

$$B_{NaCl} = 0.0902 \qquad B_{LiCl} = 0.1652$$

$$C_{NaCl} = 0.00127 \qquad C_{LiCl} = 0.0036$$

$$\Sigma m_i = 4.408 \qquad \Sigma mz = 4.408$$

The predicted values for the osmotic coefficient and water activity are the following:

$$\Phi = 1.0811 \qquad a_w = 0.92$$

Salwin–Slawson Equation

Salwin and Slawson (1959) developed a model to predict the water activity of dry mixtures by utilizing the sorption isotherms of the components:

$$a_w = \frac{\Sigma_i a_i s_i w_i}{\Sigma_i s_i w_i} \qquad (58)$$

where a_i is the initial water activity of component i used in the mixture, s_i is the sorption isotherm slope of component i at the mixture temperature, and w_i is the dry weight of component i. The isotherms can be approximated by a nonlinear equation if greater accuracy is desired (Karel, 1975a). The model is based on the assumption that no significant interactions occur to alter the isotherms of the ingredients (Bone, 1987). The water activity of one component will increase while the a_w of the other decreases in a binary mixture. The equilibrium relative humidity is based on the gain and loss of water by each component of the mixture. Figure 3.1 illustrates the principle considered by Salwin and Slawson (1959).

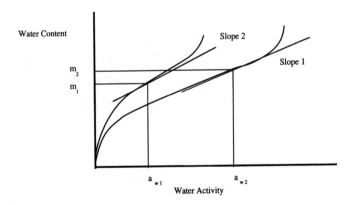

Figure 3.1. Graphical representation of Salwin and Slawson assumptions for water activity prediction. (Adapted from Salwin and Slawson, 1959).

The UNIQUAC and UNIFAC Models

The universal quasichemical equation (UNIQUAC) and the UNIQUAC Functional Group Activity Coefficients Equation (UNIFAC) use the excess Gibbs energy function to approach the prediction of thermodynamic properties of complex mixtures (Abrams and Prausnitz, 1975; Fredenslund et al., 1975; Le Maguer, 1987).

The UNIQUAC model applies to mixtures whose molecules differ appreciably in size and shape (solutes and solvent) and contains two adjustable parameters per binary interaction as established by Abrams and Prausnitz (1975). The model is a generalized form of well-known relations for the excess Gibbs energy such as Flory–Huggins, Wilson, NRTL, van Laar, Scatchard–Hamer, Margules, and Scatchard–Hildebrand (Abrams and Prausnitz, 1975). The following set of equations was reported by Abrams and Prausnitz (1975) as the UNIQUAC model:

$$\frac{G^{E++}}{RT} = \frac{G^E(\text{comb})}{RT} + \frac{G^E(\text{residual})}{RT} \tag{59}$$

where:

$$\frac{G^E(\text{comb})}{RT} = \Sigma_i x_i \ln \frac{\varphi_i}{x_i} + \frac{Z}{2} \Sigma_i q_i x_i \ln \frac{\Theta_i}{\varphi_i} \tag{60}$$

$$\frac{G^E(\text{residual})}{RT} = -\Sigma_i q_i x_i \ln\left(\Sigma_j \Theta_j \tau_{ji}\right) \tag{61}$$

$$\tau_{ji} = \exp\left(-\left(U_{ji} - U_{ii}\right)/RT\right) \tag{62}$$

$$\Theta_i = q_i x_i / \Sigma_j q_j x_j \tag{63}$$

$$\varphi_i = r_i x_i / \Sigma_j r_j x_j \tag{64}$$

where x_i is the mole fraction of component i, R is the universal gas constant, T is the temperature, q_i is the pure component area parameter, r_i is the pure component volume parameter, z is a constant set to 10, and U_{ji} is the UNIQUAC interaction parameter.

The UNIFAC model is an extension of Eq. (59) obtained by adding Debye–Hückel and Brϕnsted–Guggenheim term expressions (Christensen et al., 1983):

$$GE^* (D - H) = RT\{-\Sigma_k x_k M_k (4A/b^3)[\ln(1 + bI^{0.5})$$
$$- bI^{0.5} + b^2 I/2]\} \tag{65}$$

$$A = cd^{0.5}/(DT)^{1.5} \tag{66}$$

$$GE^*(B - G) = RT\{-\Sigma_k x_k M_k \, \Sigma_c \Sigma_a (\beta_{ca} m_c m_a/T)\} \tag{67}$$

where M is the molecular weight of component k, b is an adjustable parameter, I is the ionic strength, D is the dielectric constant of the solvent in the mixture, β_{ca} is an interaction coefficient for the ion pair c–a, c represents the cation, "a" represents the anion, and m is the molality. Abrams and Prausnitz (1975) and Fredenslund et al. (1975) listed the size and surface parameters and the binary parameters for several binary mixtures. The most relevant aspects of these two models is that they take into account the interaction effects of different groups within the molecules in the mixture.

The activity coefficient, γ_w, can be evaluated using the UNIQUAC or UNIFAC model as follows:

$$\ln \gamma_w = \frac{G^{E++}}{RT} \tag{68}$$

and the water activity can be evaluated by substituting for γ_w in Eq. (23).

3.3.3 Measurement of Water Activity

Water activity can be estimated by measuring the vapor pressure, osmotic pressure, freezing point depression, boiling point elevation, psychrometric evaluations (dew point and wet bulb depression), suction potential, or by using the isopiestic method, bithermal equilibrium, electric hygrometers, and hair hygrometers (Van den Berg, 1985; Leung, 1986).

3.3.3.1 Vapor Pressure

Water activity is expressed as the ratio of partial pressure of the water in food to the vapor pressure of pure water having the same temperature as the food. Therefore, measuring the vapor pressure of water in a food system is the most direct measure of a_w. Figure 3.2 is a diagram of a vapor pressure manometer for water activity determination. The sample is evacuated and allowed to equilibrate. The vapor pressure is measured by the manometer or a transducer. This method can be affected by the sample size, equilibration time, temperature, and volume. The method is not suitable for biological materials with active respiration or that contain large amounts of volatiles (Leung, 1986).

3.3.3.2 Freezing Point Depression and Boiling Point Elevation

Ferro-Fontán and Chirife (1981) discussed the evaluation of water activity from freezing and boiling point data.

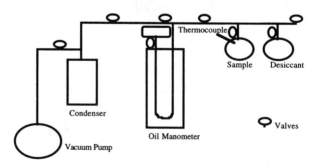

Figure 3.2.   Vapor pressure manometer.

They have reported the following equations to calculate the a_w:

Freezing point depression:

$$-\log a_w = 0.004207\Delta T_f + 2.1E\text{-}6\Delta T_f^2 \qquad (69)$$

where ΔT_f is the depression in the freezing temperature of water.

Boiling point elevation:

$$-\log a_w = 0.01526\Delta T_b - 4.862E\text{-}5\Delta T_b^2 \qquad (70)$$

where ΔT_b is the elevation in the boiling temperature of water.

This method is accurate for liquids at a high water activity range but is unsuitable for solid foods (Leung, 1986).

3.3.3.3 Osmotic Pressure

Water activity can be related to the osmotic pressure (π) of a solution by the following expression:

$$\pi = RT/V_w \ln(a_w) \qquad (71)$$

where V_w is the molar volume of water in solution, R is the universal gas constant, and T is the absolute temperature. Osmotic pressure is defined as the mechanical pressure required to prevent a net flow of solvent (water in our case) across a semi-permeable membrane. For an ideal solution, Eq. (71) can be redefined as:

$$\pi = -RT/V_w \ln(X_w) \qquad (72)$$

where X_w is the mole fraction of water in the solution. For nonideal solutions, the osmotic pressure expression can be written as:

$$\pi = -RT\phi v m_b/(m_w V_w) \qquad (73)$$

where v is the number of moles of ions formed from one mole of electrolyte, m_w and m_b are the molar concentrations of water and the solute, respectively, and ϕ is the osmotic coefficient defined as:

$$\phi = -m_w \ln(a_w)/vm_b \qquad (74)$$

3.3.3.4 Dew Point Hygrometer

Vapor pressure can be determined from the dew point of an air–water mixture. The temperature at which the dew point occurs can be determined by observing condensation on a smooth, cooled surface such as a mirror. The dew point temperature can be related to the vapor pressure by the psychrometric charts. Various pieces of equipment are available for determining dew point temperature. The formation of the dew is detected photoelectrically as shown in Figure 3.3. This method is accurate to 0.003 a_w units in the range of 0.72 to 1.00 (Troller and Christian, 1978; Leung, 1986).

3.3.3.5 Thermocouple Psychrometer

Water activity is measured based on wet bulb depression. A thermocouple is cooled in the chamber where the sample is equilibrated, and the water is condensed over the thermocouple. Once the thermometer is wet, the water is

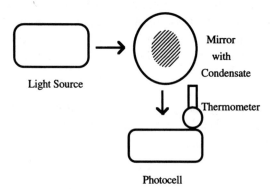

Light Source

Mirror with Condensate

Thermometer

Photocell

Figure 3.3.   Diagram of a dew point determination device.

allowed to evaporate, causing a decrease in temperature. The drop in temperature is related to the rate of water evaporation from the surface of the thermometer, which is a function of the relative humidity in equilibrium with the sample.

3.3.3.6 Isopiestic Methods and Graphic Interpolation

The isopiestic method consists of equilibrating both a sample and a reference material in an evacuated dessicator for 24 h at 25°C. The moisture content of the reference material is then determined and the water activity obtained from its sorption isotherm. Because the sample was in equilibrium with the reference material, the water activity of both is the same (Troller and Christian, 1978; Leung, 1986).

Microcrystalline cellulose and potato starch can be used as reference materials to determine water activity of food products (Wolf et al., 1984). Figure 3.4 shows the standard sorption isotherms of microcrystalline cellulose and potato starch at 25°C (Wolf et al., 1984).

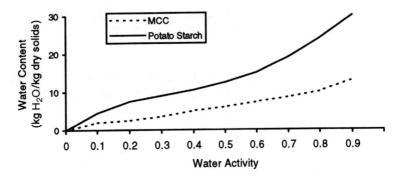

Figure 3.4.   Standard sorption isotherm of microcrystalline cellulose and potato starch at 25°C (Wolf et al., 1984).

The use of sulfuric acid solutions has been reported when the graphical interpolation method is considered. Saturated or unsaturated salt solutions may also be used to obtain prescribed water activity levels. Table 3.3 lists several salts and their water activity values at 25°C. The weight gain and loss of the sample is related to the water activity of the acid or salt solution. The a_w is estimated when no weight changes are observed (Troller and Christian, 1978; Leung, 1986).

3.3.3.7 Electric and Hair Hygrometers

Most hygrometers are electrical wires coated with hygroscopic salt or a sulfonated polystyrene gel in which conductance or capacitance changes as the coating absorbs moisture from the sample. The principal disadvantage of this type of instrument is the tendency of the hygroscopic salt to become contaminated with polar compounds, which results in reduced or erratic water activity readings.

The hair hygrometer is based on the stretching of a fiber when it is exposed to high water activity. It is less sensitive than other instruments ($\pm$ 0.03 a_w) and the main disadvantages of this type of hygrometer are the time delay in reaching equilibrium and the tendency to hysteresis (Troller and Christian, 1978).

Table 3.3.   Water activity values of different salt solutions at 25°C, as reviewed by Troller and Christian (1978).

Salt	a_w	Salt	a_w
LiCl	0.11–0.15	$KC_2H_3O_2$	0.20–0.23
$MgCl \cdot 6H_2O$	0.33	K_2CO_3	0.44
$Mg(NO_3)_2 \cdot 6H_2O$	0.52–0.55	NaCl	0.75
$CdCl_2$	0.82	K_2CrO_4	0.88
KNO_3	0.93–0.94	K_2SO_4	0.97

3.3.3.8 Water Potential

The concept of water potential has been widely used by soil and plant scientists. It can be defined as the difference between the chemical potential of the water in the system and that of pure water at the same temperature.

$$(\partial\mu/\partial P)_T = V_w \tag{75}$$

then

$$\Delta\mu = V_w\Delta P \tag{76}$$

and the water potential (χ_w) can be expressed as:

$$\psi_w = \Delta P = \Delta\mu/V_w \tag{77}$$

ΔP can be considered the suction required on pure water to reduce its activity to that of the food at the same temperature. The ψ_w can then be related to water activity using Eq. (4):

$$\psi_w = RT/V_w \ln(a_w) \tag{78}$$

The instrument to measure ψ_w consists of a porous cup or membrane that is permeable to water and solutes but not to air and macromolecules. The cup is filled with water and connected to a manometer or vacuum gauge so that the suction potential can be measured.

3.4 SORPTION PHENOMENA AND SORPTION ISOTHERMS

The relationship between total moisture content and the corresponding water activity over a range of values at a constant temperature yields a moisture sorption isotherm. Sorption isotherms exist in the four main food processing areas: drying, mixing, packaging, and storage (Jowitt et al., 1981). The sorption process is not fully reversible. Adsorption isotherms are used for the observation of hygroscopic products; desorption isotherms are useful for the investigation of the drying process. Water sorption occurs first by the formation of a monolayer over the product surface and is followed by multilayer adsorption. Multilayer water adsorption consists of water uptake into pores and capillary spaces, dissolution of solutes, and finally the mechanical entrapment of water. These phases may overlap extensively and will differ among foods, depending on composition and structure (Troller and Christian, 1978).

3.4.1 Hysteresis Phenomenon

The term *hysteresis* describes the phenomenon where the adsorption and desorption paths of an isotherm are different. It has important theoretical implications such as the irreversibility of the sorption process, and it may be considered a protective mechanism against dry atmosphere, frost damage, and freezer burns (Kapsalis, 1987). Figure 3.5 shows a schematic representation of an hysteresis loop. The desorption isotherm usually lies above the adsorption isotherm in the midrange of water activity values.

The hysteresis effect can be explained as the effect of water condensed in the capillaries. The so-called *Ink Bot-*

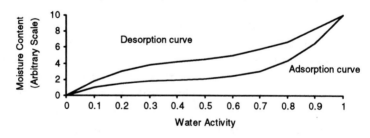

Figure 3.5. Sorption isotherms and hysteresis phenomena as discussed by Labuza (1968).

tle Theory explains the phenomenon (Labuza, 1968). It is assumed that capillaries have narrow necks and large bodies, as represented schematically in Figure 3.6. The capillary will not fill completely during adsorption until a higher water activity is reached, which corresponds to the large radius *R*. The emptying process, during desorption, is controlled by the smaller radius *r*, thus lowering the water activity (de Man, 1982).

The hysteresis in food products can be broken down into three groups (Kapsalis, 1987): (1) in high-sugar and high-pectin foods, the phenomenon is pronounced in the lower moisture content region; (2) in high-protein foods, the hysteresis begins at high water activity in the capillary condensation region and extends over the isotherm to zero water activity; and (3) in starchy foods, a large loop is reported with a maximum water activity value of 0.70 (or within the capillary condensation region).

The effect of temperature on the magnitude of hysteresis varies among foods. For some foods the phenomenon can be eliminated by increasing the temperature, whereas for others hysteresis can remain constant or increase with an increase in temperature (Kapsalis, 1987).

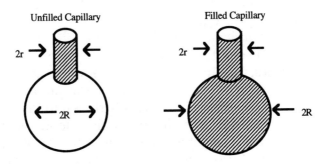

Figure 3.6.   Ink bottle theory of hysteresis as discussed by Labuza (1968).

3.4.2 Temperature Effect on Sorption

The effect of temperature is of great importance because foods are not composed of ideal mixtures, and water activity changes with temperature. As an important parameter associated with stability and the moisture sorption process, the water activity should be predicted at different conditions.

A plot of the natural logarithm of water activity, $\ln(a_w)$, against the reciprocal of the absolute temperature, $1/T$, at a constant moisture content will generally give a straight line. The slope of this line is equal to Q/R where R is the universal gas constant, and Q is the net isosteric heat of sorption. The latter is defined as the difference between the total molar enthalpy change and the molar enthalpy of vaporization:

$$d(\ln(a_w))/d(1/T) = -Q/R \qquad (79)$$

Equation (79) is known as the Clausius–Clapeyron relationship. It provides useful information on the binding energy of water molecules, as discussed in Section 3.4.4, while evaluating the heat of sorption as a function of moisture content. Figure 3.7 presents an example of a Clausius–Clapeyron representation as discussed by Vega-Mercado and Barbosa-Cánovas (1993b) for freeze-dried pineapple pulp.

Temperature affects the mobility of water molecules and the equilibrium between the vapor and the adsorbed phases. An increase in temperature, at constant water activity, causes a decrease in the amount of adsorbed water. An exception to this is shown in the case of certain sugars and low molecular weight food constituents that dissolve in water and become more hygroscopic at higher temperatures. On the other hand, the chemical and microbiological reactivity is affected by the temperature-moisture content relationship, because an increase in

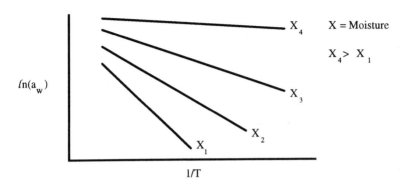

Figure 3.7.   Clausius–Clapeyron relationship for different moisture contents.(Adapted from Vega-Mercado and Barbosa-Cánovas, 1993b).

temperature results in an increase of the water activity at constant moisture content.

3.4.3 Brunauer–Emmett–Teller (BET) Equation

The relationship between total moisture content and the corresponding water activity over a range of values at constant temperature yields a moisture sorption isotherm. The main application of sorption isotherms is in drying, mixing, packaging, and storing (Jowitt et al., 1981).

The prediction of water sorption isotherms is based on the Langmuir equation:

$$\frac{a_{\mathrm{w}}}{X_{\mathrm{e}}} = \frac{k}{bX_{\mathrm{m}}} + \frac{a_{\mathrm{w}}}{X_{\mathrm{m}}} \tag{80}$$

where X_{e} is the equilibrium water content, X_{m} is the monolayer value, k is the inverse of pure water vapor pressure at the temperature of the system, and b is a constant. Unfortunately, this equation is not satisfactory for food because the heat of adsorption is not constant over the whole surface; there is a high interaction between adsorbed molecules, and the maximum amount of water adsorbed on the surface is greater than a monolayer (Labuza, 1968; De Man, 1982; Vega-Mercado and Barbosa-Cánovas, 1993a).

An effective way of estimating the contribution of adsorption at specific sites to total water binding is the use of the Brunauer–Emmett–Teller (BET) equation (Labuza, 1968, Karel, 1975a). The equation can be expressed as:

$$\frac{X_e}{X_m} = \frac{Ca_w}{(1-a_w)(1+a_w(C-1))}$$

(81)

$$C = Ke^{\frac{Q}{RT}}$$

(82)

where X_e is the adsorbed water content, X_m is the adsorbed monolayer value, a_w is the water activity, Q is the heat of sorption, T is the temperature, and K is a constant. This equation can be applied to water activities from 0.1 to 0.5 because the assumptions considered for moist materials are not entirely true (Labuza, 1968; Karel, 1975). Figure 3.8 shows a BET plot. The slope and inter-

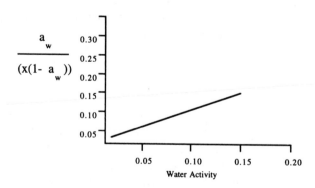

Figure 3.8.   BET plot for dehydrated orange crystals at 25°C (Karel, 1975a).

cept of $a_w/X (1 - a_w)$ versus a_w are used to determine the constant C and the monolayer value X_m:

$$\text{slope} = \frac{(C-1)}{(X_m C)} \tag{83}$$

$$\text{intercept} = \frac{1}{(X_m C)} \tag{84}$$

3.4.4 Heat of Sorption and Free Energy Change

The heat of sorption, Q, should be constant up to monolayer coverage and then suddenly decrease (Labuza, 1968; Román et al., 1982; Lima and Cal-Vidal, 1983; De Gois and Cal-Vidal, 1986; Vega-Mercado and Barbosa-Cánovas, 1993b). Figure 3.9 shows the relationship of heat of sorption and moisture content as actually observed and according to the BET theory. The heat of sorption at low moisture content is higher than the theory indicates and also falls off gradually, indicating the gradual change from monolayer water to capillary water (de

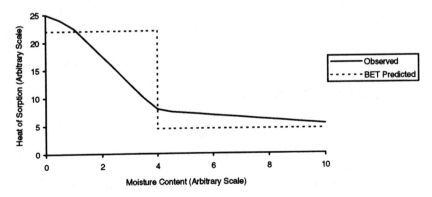

Figure 3.9. Heat of sorption and water content relationship as discussed by Labuza (1968).

Man, 1982). The main reason for the increase of water content at high values of water activity must be capillary condensation.

Chung and Pfost (1967a, 1967b), Rockland (1969), Kumar (1974), and Rockland and Nishi (1980) discussed the relationship of water content in food and the free energy change (ΔF) during the sorption phenomena:

$$\log(-\Delta F) = nX_e + k \qquad (85)$$

$$\Delta F = RT \ln(a_w) \qquad (86)$$

where n and k are constants. Chung and Pfost (1967a) defined ΔF as the work done by the attractive forces during sorption.

3.4.5 Empirical and Theoretical Models of Sorption Isotherms

Van den Berg and Bruin (1981) have reported more than 200 isotherm equations proposed for biological materials. The equations vary from empirical models with two or three fitting parameters, to rigorously derived thermodynamic models, to equations derived based on the BET model (Boquet et al., 1980). Sorption isotherms are useful in the prediction of food product shelf life. The following discussion covers only the most relevant equations reported to describe sorption isotherms of foods.

3.4.5.1 Henderson Model

In 1952, Henderson reported an equation that represents conventional equilibrium moisture data:

$$a_w = 1 - \exp(-k'X_e^n) \qquad (87)$$

where n and k' are the model constants. Equation (87) can be expressed considering the temperature effect as follows (Henderson, 1952):

$$a_w = 1 - \exp(-k''TX_e^n) \qquad (88)$$

where k'' is another constant of the model. The equation makes possible the extrapolation of limited sorption data and the determination of the temperature effect on sorption (Henderson, 1952). Lima and Cal-Vidal (1983) reported the use of the Henderson model for fitting sorption data of freeze-dried banana and Vega-Mercado and Barbosa-Cánovas (1993a) represented the sorption data of freeze-dried pineapple pulp.

3.4.5.2 Iglesias-Chirife Model

Iglesias and Chirife (1978) worked on an empirical model for describing the water sorption behavior of fruits and related high-sugar products. The model is expressed as:

$$\ln(X_e + (X_e^2 + X_{in})^{0.5}) = ba_w + p \qquad (89)$$

where X_{in} is the equilibrium moisture content at a_w equal to 0.5, and b and p are the fitting parameters of the model.

3.4.5.3 Guggenheim-Anderson-de Boer (GAB) Model

The GAB model is an extension of the BET equation taking under consideration the modified properties of adsorbed water in the multilayer region (Kapsalis, 1987). The model is reported to be the best for fitting sorption isotherm data to the majority of food products (Kapsalis, 1987; Vega-Mercado and Barbosa-Cánovas, 1993a). The GAB model is very useful to predict data up to water activity levels of approximately 0.9, and it gives a better fit than the BET equation over a wide range of moisture contents. The GAB model is expressed as:

$$\frac{X_e}{X_m} = \frac{CKa_w}{(1 - Ka_w)(1 - Ka_w + CKa_w)} \qquad (90)$$

$$C = c \exp\left(\frac{H_m - H_n}{RT} \right) \tag{91}$$

$$K = k \exp\left(\frac{H_p - H_n}{RT} \right) \tag{92}$$

where c and k are entropic accommodation factors, H_m is molar sorption enthalpy of the monolayer, H_n is molar sorption enthalpy of the multilayer, and H_p is molar enthalpy of the evaporation of liquid water. The monolayer value obtained by the GAB model is generally higher than that obtained by the BET model.

3.4.5.4 Smith Model

The Smith equation (Smith, 1947) has been usefully applied to a water activity of 0.3 to 0.5. The equation is expressed as:

$$m = A - B \ln(1 - a_w) \tag{93}$$

where A and B are the model constants. Equation (93) is considered by Lang and Steinberg (1981) to predict the effect of a nonsolute (i.e., starch, casein, soy flour) on water activity. The model applies to adsorption data above water activity of 0.5.

3.4.6 Sorption Isotherm and Water Activity Standards

The use of a reference material with defined and stable sorption properties has been suggested to establish a standard method for sorption measurements. Wolf et al. (1984) established the following guidelines for the selection of the material: (1) stable sorption behavior over several adsorption and desorption cycles and at extreme ambient temperatures when shipped or stored; (2) absence of hysteresis; (3) rapid rate of equilibration; (4) sorption isotherm sigmoid in shape; (5) homogeneous structure and component distribution; (6) determination of water content in an unequivocal way; and (7) accessibility and ease of handling.

Microcrystalline cellulose (MCC), as a reference material for sorption isotherms, fulfills the above criteria (Wolf et al., 1984). Figure 3.10 shows the adsorption isotherm of MCC at 25°C. The product is stable in its crystalline structure over a range of –18°C to 80°C.

Saturated salt solutions are being used as a sorbate source. Rockland (1960), Karel (1975), Wolf et al. (1984), and Kitic et al. (1986) reported on some of the most common salt solutions available for sorption experiments. Saturated salt solutions have the advantage of maintaining a constant relative humidity as long as the amount of salt present is above the saturation level. A list of salts used to control water activity is shown in Table 3.4.

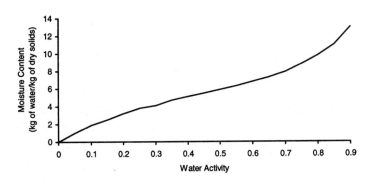

Figure 3.10.   Sorption isotherm for MCC at 25°C (Wolf et al., 1984).

Table 3.4.   Saturated salt solutions used for controlling water activity.

Salt	Solubility (g/100g of H_2O)	Water Activity (a)	(b)	(c)	(d)
$LiCl.H_2O$	100	0.11	—	0.11	0.11
$MgCl.6H_2O$	300	0.32	—	0.33	0.32
K_2CO_3	150	0.43	—	0.43	0.43
$NaBr.2H_2O$	120	0.58	0.56	0.56	0.57
$NaCl$	40	0.75	0.75	0.75	0.75
KCl	50	0.85	0.84	0.86	0.84
K_2SO_4	20	0.97	0.98	0.97	—
K_2CrO_4	200	—	—	0.87	—

Note: (a) Karel (1975a), (b) Kitic *et al.* (1986), (c) Rockland (1960) and (d) Wolf *et al.* (1984)

3.5 FOOD STABILITY

Generally microbial spoilage and chemical deterioration of foods are related to water activity. The relationship may be summarized by Figure 3.11, where all the factors related to the stability of foods are shown. The following stability aspects are discussed in detail with respect to the water activity and water content of food products: (1) microbial changes; (2) enzymatic and nonenzymatic reactions; (3) physical and structural changes; and (4) nutrient, aroma, and flavor destruction.

3.5.1 Microbial Spoilage

Living cells are sensitive to the water status in their surrounding environment. The ability of microorganisms to grow and produce toxins is related to the water activity of the media. Beuchat (1983) and Scorza et al. (1981) have reported minimal water activities required for growth of a number of microorganisms (Table 3.5). Optimal and reduced water activity growth curves of bacteria are shown in Figure 3.12. Karel (1975) and Beuchat (1983)

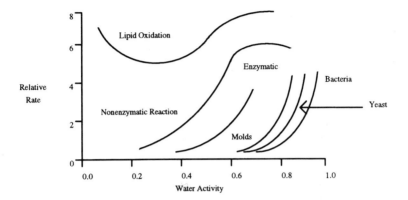

Figure 3.11.   Water activity and stability diagram as presented by Labuza et al. (1972).

Table 3.5.   Minimal water activity for growth of microorganisms.

Organism	Classification	a_w
Aspergillus flavus	Mold	0.78
Aspergillus parasiticus	Mold	0.82
Penicillium citrinum	Mold	0.80
Aspergillus ochraceus	Mold	0.83
Clostridium botulinum Type E	Bacteria	0.98
Pseudomonas	Bacteria	0.98
Flavobacterium	Bacteria	0.96
Klebsiella	Bacteria	0.96
Lactobacillus	Bacteria	0.93
Micrococcus	Bacteria	0.90
Staphylococcus (aerobic)	Bacteria	0.86
Rhodattorula	Yeast	0.92
Saccharomyces	Yeast	0.90
Candida	Yeast	0.88

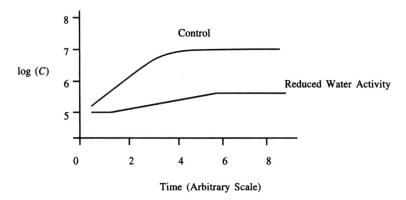

Figure 3.12.   Growth curve of bacteria (Troller, 1987).

have reported that minimum water activities for the production of toxins are often higher than those for microbial
growth. The effect of reduced water activity can be
observed in the logarithmic phase, where the time
required for individual cells to reproduce is extended.

The stationary phase is also affected by the water
activity. Troller (1987) has reported the growth curves of
Staphylococcus aureus at various water activity levels, as
shown in Figure 3.13, as well as the effect of water activity for each of these growth phases. Other factors combined with water activity that affect the inhibition of
microorganisms are pH, oxygen, preservatives, temperature, and radiation.

Karel (1975a) has investigated the sensitivity of
microorganisms to heat, light, and chemicals as a function of water activity. It was found, in all three cases, the
higher the water activity the higher the sentivity.

3.5.2 Enzymatic Reactions

Enzymatic activity in food products is inhibited at water
activities lower than 0.75. Short heat treatments, such as

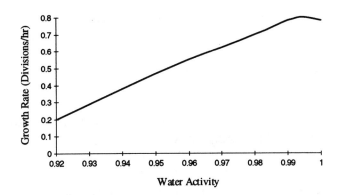

Figure 3.13. *Staphylococcus aureus* **growth curve as a function of water activity (Troller, 1987).**

blanching of raw vegetables, are used to inactivate enzymes. Improvement in sensory properties of foods may be achieved by enzymatic activity, such as the production of glucose from starch while processing domestic animal feed. In other cases, enzymatic activity is not desirable because it affects the amount of nutrients in food, such as in the hydrolysis of lecithin by phospholipase.

The reduction in enzymatic activity is not related to structural changes of the enzyme in all cases. Factors such as ionic strength and competition for active sites by nucleophilic compounds, such as methanol, ethanol, and glycerol, explain the stability of the enzymatic complex in food systems.

Dark pigment formation, production of gases, and the reduction in fruit volumes are reported during enzymatic browning of food products (Monsalve-González et al., 1993b). Oxidoreductases are responsible for these types of reactions and are known as phenoloxidase, tyrosinase, cathecolase, phenolase, polyphenol oxidase, and polyphenolase. The most common substrates related to enzymatic browning are unsaturated compounds such as monophenols and o-diphenols. Polyphenol oxidase (PPO) is responsible for browning in sliced fruits and vegetables when the slices are exposed to air. The enzyme promotes the oxidation of 2-o-diphenol to 2-o-quinone. A typical scheme for a PPO reaction is shown in Figure 3.14.

The concentration of the PPO enzyme is the limiting factor of the enzymatic reaction in the food system. There is an abundance of the substrate, so changes in its concentration have no effect on the reaction rate. The enzymatic reaction rate can be expressed as:

$$-dS/dt = a \qquad (94)$$

where S is the substrate concentration, t is time, and a is the activity of the enzyme, defined as the rate at which a specific quantity of the enzyme will convert a substrate to product (Toledo, 1991). Equation (94) is valid when there are no inhibition effects on the enzymatic reaction. However, most enzymatic reactions proceed in a curvilinear pattern that can be expressed as:

$$-dS/dt = a - B \qquad (95)$$

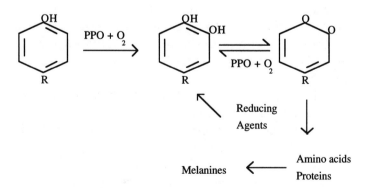

Figure 3.14. Typical reaction mechanism for enzymatic browning by PPO.

where B represents the inhibition of the enzyme. Assuming B is proportional to the product formed, Eq. (95) can be redefined as:

$$-dS/dt = a - k_i P \qquad (96)$$

where P is the product concentration and k_i is the proportionality constant. When the initial enzyme activity is quite high relative to the substrate concentration, the fraction converted may be a linear function of time. The influence of product inhibition becomes significant once the product accumulates, the substrate concentration drops, and the substrate conversion occurs exponentially (Toledo, 1991).

3.5.3 Nonenzymatic Reactions

The most common nonenzymatic reaction in low-moisture food products is the browning reaction. Two types of browning can be defined: caramelization and Maillard reactions. The browning in many foods may be aesthetically desirable or undesirable because of the colors, aromas, and flavors that can develop. In many cases, avoiding the browning reaction will prevent the loss of essential nutrients such as amino acids.

Caramelization is promoted by the direct heating of carbohydrates (sugars and sugar syrups) which causes anomeric shifts, ring size alterations, breakage of glycosidic bonds, and the formation of new bonds (Whistler and Daniel, 1985). The introduction of double bonds into sugar rings produces intermediates to unsaturated rings, which absorb light and produce color. Other factors that influence the browning reaction are pH and the presence of oxygen, metals, phosphates, and sulfur dioxide.

The Maillard reaction occurs via a series of complex defined reactions between reducing sugars and amino groups in amino acids and proteins (Troller and Christian, 1978). These reactions result in the loss of nutritive value, formation of brown pigment, as well as the formation of *off-flavors*. The reaction has a maximum rate in the low moisture range; for example, the maximal rate in fruits ranges from 0.65 to 0.75 water activity (Troller and Christian, 1978).

The Maillard reaction involves a number of steps. Figure 3.15 outlines the mechanism of brown pigment or melanoidin formation (de Man, 1982). de Man (1982) identified five basic steps involved in the process:

1. The production of an N-substituted glycosylamine from an aldose or ketose reacting with a primary amino group of an amino acid, peptide, or protein.

2. Rearrangement of the glycosylamine by an Amadori rearrangement reaction to yield an aldoseamine or ketoseamine.

3. A second rearrangement of the ketoseamine with a second mole of aldose to result in the formation of a diketoseamine, or the reaction of an aldoseamine with a second mole of amino to yield a diamino sugar.

4. Degradation of the amino sugar with loss of one or more molecules of water to give amino or non-amino compounds.

5. Condensation of the compounds formed in step 4 with each other or with amino compounds, resulting in the formation of brown pigments and polymers.

The formation of glycosylamines from the reaction of amino groups and sugars is reversible (de Man, 1982), and the equilibrium concentration is highly dependent on the moisture level. The rate of browning is high at a low water content. This fact explains the ease of browning in dried and concentrated foods.

Temperature has some effect on the browning rate. A two to threefold increase for each 10°C rise in temperature is reported on model systems. In the case of foods

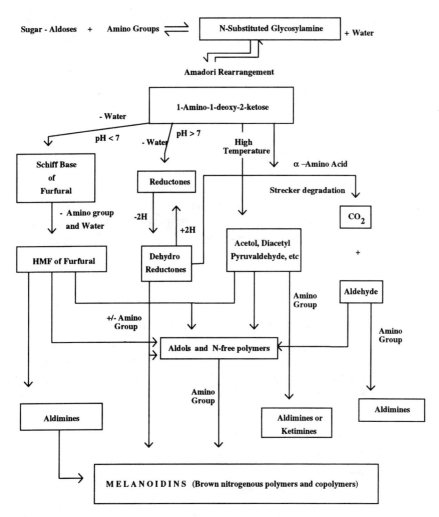

**Figure 3.15.   Mechanism of melanoidin formation during the nonen-
zymatic browning reaction (de Man, 1982).**

containing fructose, the increase of browning may be five to tenfold for each 10°C rise. The browning reaction is retarded by decreasing the pH in the range of pH 8 to 3. The reaction can be said to be self-inhibitory because of the decrease in pH with the loss of the basic amino group. The effect of pH on the browning reaction is highly dependent on the moisture content.

3.5.4 Lipid Oxidation

The oxidation reactions in foods are related to the presence of metals, which act as catalysts for free radicals. Figure 3.16 shows a general pathway for lipid oxidation as described by Nawar (1985). The reaction rate is a function of water activity as shown in Figure 3.11. At low a_w values, the reaction takes place slowly because of limited mobility and availability of free radicals. This situation coincides with the Brunauer–Emmett–Teller (BET) determined monolayer value (Karel, 1975a). At high a_w values, over the monolayer amount, the reaction rate increases by the dissolution of metal ions, which promotes the formation of free radicals. This type of reaction is prevented by different methods (Troller and Christian, 1978; Villar and Silvera, 1987):

- Reduction of oxygen content and its diffusion.
- Use of antioxidants such as BHA, BHT, PG, and tocophenols
- Use of synergistic agents such as citric and phosphoric acid
- Use of chelating agents such as EDTA, citric, or malic acid

3.5.5 Physical and Structural Phenomena

Geometry and structural changes on food materials during drying affects the mass transport properties and the quality of the product (Karel, 1975b; Okos et al., 1992).

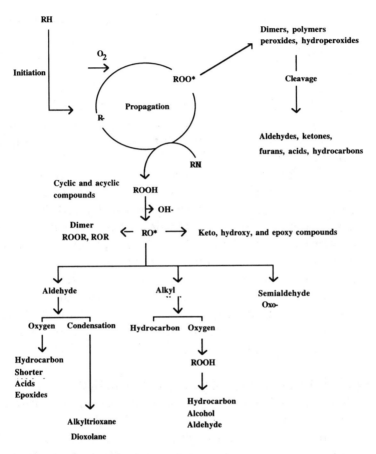

Figure 3.16.   General scheme of lipid oxidation (Nawar, 1985).

The most common defects of dehydrated foods are a tough, woody texture; slow or incomplete rehydration; and loss of the juiciness typical of fresh foods. The crystallization of polysaccharides such as starches is promoted by the removal of water. This phenomenon, in addition to the loss of cellular integrity, explains some of the changes in plant materials. Animal tissue tenderness losses are explained based on the aggregation of muscle proteins.

Shrinkage is the most common change during dehydration and occurs in the early drying stages (Okos et al., 1992). The shrinkage and toughening in food products result from the collapse of the structure which is related to the glass transition temperature (T_g). Food dehydration in the vicinity of or below the glass transition temperature (T_g) may prevent the shrinkage or structure collapse (Karel et al., 1993). Unfortunately, fruits and vegetables have high initial moisture content, therefore very low T_g values, which makes it very difficult to avoid the shrinkage. Typical air drying operations are performed at temperatures much higher than their T_gs, which leads to extensive shrinkage and collapse. Food dehydration using low temperatures (i.e., freeze drying) results in increased porosity and reduced shrinkage but it is expensive because of the extended dehydration time. In the following section we discuss briefly the *glass transition temperature* concept and its application in food dehydration.

3.5.5.1 Glass Transition Temperature

The approach known as food polymer science proposed by Slade and Levine (1991, 1993) has provided an additional interpretative and experimental framework for the study of food systems. Key elements of this approach include (Slade and Levine, 1991):

1. The behavior of foods as classical polymer systems; that behavior is governed by dynamics rather than energetics.

2. A characteristic temperature, T_g, at which the glass–rubber transition occurs, used as a physicochemical parameter that can determine processability, properties, quality, stability, and safety of foods.

3. The central role of water as an ubiquitous plasticizer of natural and fabricated amorphous foods.

4. The effect of water as a plasticizer on T_g, which results in a non-Arrhenius diffusion-limited behavior of amorphous polymeric, oligomeric, and monomeric foods at $T > T_g$.

5. The significance of non-equilibrium glass solid and rubbery solid state in most food products and processes, and the effects on time-dependent structural and mechanical properties related to quality and storage stability.

The term *glassy* is used to identify a material that display brittleness, strength, transparency, and progressive softening when heated (Allen, 1993). Thus, a definition of glass formation could be, as stated by Allen (1993):

> "a substance in the glassy state is a material formed by cooling from the normal liquid state, which shows no discontinuous change at any temperature, but has become rigid through a progressive increase in viscosity."

A curve of viscosity (η) versus temperature does not present a sharp transition from liquid to glass. The glass transition temperature is defined as the intersection of the linear projections from both glass and liquid as presented in Figure 3.17. Some glass-forming compounds are organic polymers, glycerol and other hydrogen-bonded systems, and carbohydrate–water and protein–water systems of various composition.

Sharp state changes of first-order transitions are characterized by discontinuities in the primary thermodynamic properties such as volume, enthalpy (H), and free energy (G). Glass-forming liquids do not present those discontinuities at the T_g temperature, which is known as second order transition. Figure 3.18 presents both first- and second-order transitions for nonglass and glass-forming liquids.

The second-order transition of amorphous food occurs when a glass is transformed into a rubber during heating over the glass transition temperature T_g. The

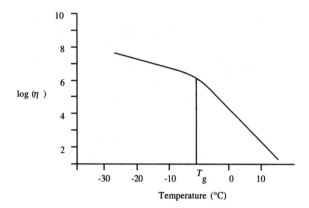

Figure 3.17.   Glass transition temperature as a function of viscosity (Adapted from Allen, 1993).

amorphous state is metastable. In foods containing amorphous components, the state of the food may then be defined at a given time by the transport and relaxation phenomena (Slade and Levine, 1991; Roos, 1992). The critical effect of plasticization, leading to increased free volume and mobility in the dynamically strained glass by water at T_g, is a key factor of collapse and its mechanism

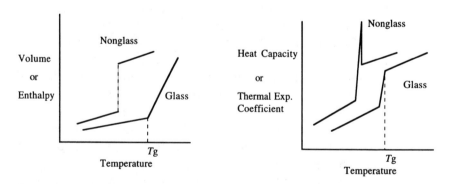

Figure 3.18.   First- and second-order transitions (Adapted from Allen, 1993).

as stated by Slade and Levine (1991). The viscoelastic behavior of amorphous polymeric networks (i.e., gluten and elastin) and the gelation of starches have been described in terms of the T_g concept.

Slade and Levine (1991) discussed the importance of the free-volume theory in the description of the second-order transition in amorphous food materials. The free volume theory is based on the fact that T_g can also be defined as the temperature at which the thermal expansion coefficient of the material changes. The free volume (V_f) can be defined as (Roos, 1992):

$$V_f = V - V_o(T) \qquad (97)$$

where V is the macroscopic volume and V_o is the volume occupied by the molecules at temperature T. Low molecular weight plasticizers, such as water, lead to much greater mobility because of increased free volume and decreased local viscosity as moisture content increases from that of a dry solute to a solution. Figure 3.19 shows the relationship between T_g and water content of gelatinized starch as discussed by Slade and Levine (1991) and Van den Berg (1985).

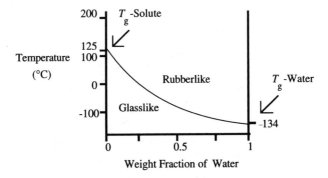

Figure 3.19. Relationship of T_g and water content in a gelatinized starch mixture (Van den Berg, 1985).

It has been proposed that the effects of moisture and temperature on chemical changes during the processing and storage of foods are related to the physical state of the product and specifically to temperatures above the glass transition condition (Karel et al., 1993). Figure 3.20 presents the relationship of water activity and T_g as discussed by Karel et al. (1993). The lower the water activity in a food the higher the glass transition temperature of the product.

The concepts of *water dynamics* and *glass dynamics* are summarized on a dynamic map shown in Figure 3.21. The equilibrium regions represented on the diagram are not time dependent, and they are represented in two dimensions by temperature and composition at atmospheric pressure. The nonequilibrium conditions require the third dimension of time, expressed as t/τ where τ is a relaxation time (Slade and Levine, 1991).

Water dynamics, which focuses on the mobility and availability of the plasticizing diluent (water), is used as an approach to understand how to control the mobility of the diluent in glass-forming food systems that are inherently mobile, unstable, and reactive at temperatures

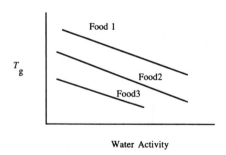

Water Activity

Figure 3.20.   Relationship between water activity and glass transition temperature (Adapted from Karel et al., 1993).

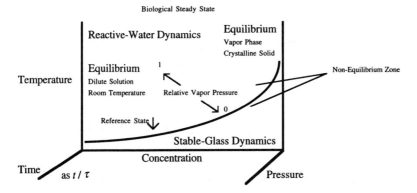

Figure 3.21. Dynamic map for a food system (Adapted from Slade and Levine, 1991).

above T_g and moisture content above W_g (amount of water over the glass transition condition).

Glass dynamics deals with the time–temperature dependence of relationships among composition, structure, thermomechanical properties, and functional behavior. It focuses on glass-forming solids in an aqueous food system, the T_g value for the aqueous glass, and the effect of the glass transition on processing and process control.

The Williams–Landel–Ferry equation (Williams et al., 1955) was obtained empirically to describe time and temperature dependence of mechanical properties (a_T) of amorphous materials.

$$a_T = \frac{C_1 + \left(T - T_g\right)}{C_2\left(T - T_g\right)} \tag{98}$$

or expressed for the vicinity of T_g:

$$\log \frac{\eta}{\eta_g} = \frac{-17.44 + \left(T - T_g\right)}{51.6\left(T - T_g\right)} \tag{99}$$

where C_1 and C_2 are constants, T is the temperature, T_g is the glass transition temperature, η is viscosity and ηg is the viscosity at T_g. The equation applies for $T_g < T < T_g + 100°C$, while an Arrhenius type dependence is considered when $T < T_g$ and $T < T_g + 100°C$ (Roos, 1992).

A gelatinized material is completely permeable to the diffusion of dispersed dyes and large molecules at $T_g < T < T_m$ (melting temperature), while the diffusion is inhibited when $T < T_g$. Figure 3.22 illustrates the changes in viscoelastic properties as a function of temperature on gel: (1) vitrified glass when $T < T_g$; (2) glass transition to a leathery region when $T = T_g$; (3, 4) at $T_g < T < T_m$ rubbery plateau to rubbery flow; and (5) at $T > T_m$, viscous liquid flow (Slade and Levine, 1991).

Karel et al. (1993) defines a generic state diagram for a water-soluble food component, as presented in Figure 3.23, where important food quality modifying events such as collapse, crystallization, and stickiness are shown. A similar diagram may be used to describe the path of food processes such as dehydration, freezing,

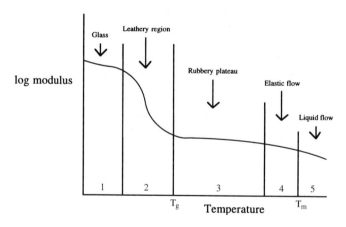

Figure 3.22.   Changes in viscoeleastic properties of synthetic partially crystalline polymers (Adapted from Slade and Levine, 1991).

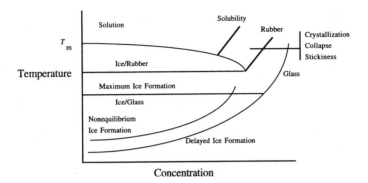

Figure 3.23.   State diagram showing stability of glassy state (Adapted from Karel et al., 1993).

cooling, and heating as presented in Figure 3.24. The packing or conformational rearrangement of the molecules in a nonequilibrium state is derived from the total change of the Gibbs energy. A decrease in enthalpy and volume will be observed when sub-T_g annealing is considered. Also, an increase in density, tensile and flexural yield stress and elastic modulus, decreased impact strength, fracture energy, ultimate elongation, and creep

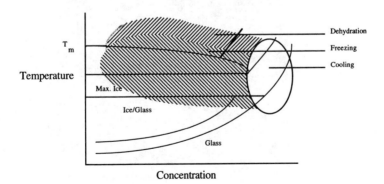

Figure 3.24.   Food processes location in the state diagram (Adapted from Karel et al., 1993).

rate result from the conformational rearrangement (Tant and Wilkes, 1981). The change in molecular mobility and free volume leads to more brittle materials. Many dried foods becomes glassy after drying because of collapse, stickiness, caking, and crystallization occurring when some of the components are in the vicinity of their T_g temperature.

Enzymatic reactions are essentially inhibited in a partially glassy solid at $T < T_g$, but the reaction can proceed when the rubbery state $(T < T_g)$ is attained (Slade and Levine, 1991). Rates of enzymatic reaction would be an exponential function of ΔT between the freezing temperature (T_f) and T_g' (or the threshold temperature for the onset of enzymatic activity in a frozen system). The same principle applies to chemical reactions when the reac-

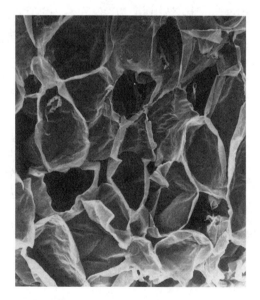

Figure 3.25.   SEM photo of freeze-dried apple treated by combined methods (3 h).

tants are prevented from moving within the glassy product and no reaction takes place.

3.5.5.2 Techniques to Study Food Structure

Physical and chemical changes of foods during drying can be studied using scanning electron microscopy (SEM), nuclear magnetic resonance (NMR), and image analysis techniques. Figures 3.25 and 3.26 show SEM photos of freeze-dried apple tissues treated by combined methods before drying. The changes on the organic matrix (cellular walls) are noticeable. The use of SEM provides a good tool to study this type of phenomena as well as other types of physical and chemical changes during the drying of food materials.

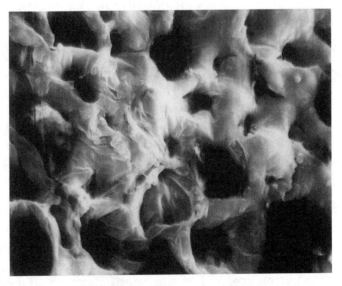

Figure 3.26.   SEM photo of freeze-dried apple treated by combined methods (10 h).

3.5.6 Destruction of Nutrients, Aromas, and Flavor

Compounds that impart aromas and flavors are naturally occurring chemicals in foods with relatively high vapor pressures (Karel, 1975c; Bruin and Luyden, 1980). A consequence of drying is the evaporation and loss of these compounds that occurs during the heating of the product and the removal of water. Isothermal drying of gelled slabs demonstrated that aroma retention increases with an increase in initial concentration of dissolved solids, thickness of the slab, mass transfer coefficient in the gas phase, slab temperature, and the decrease in the relative humidity of drying air.

The retention of volatiles during freeze drying has been explained by the entrapment of the compounds within the dry food matrix (Karel, 1975c). The decrease in the diffusion coefficient of volatiles appears to be the explanation for a preferential retention of this type of compound as compared to water. The concept of microregion permeability has been discussed by Karel (1975c) as an explanation of the above phenomenon. Retention of volatiles in freeze-dried materials is not caused by an adsorption mechanism in the dry layer. Thus, after the freeze drying the volatiles are strongly held in the material in small, localized, and uniformly distributed units in the dry matrix, known as microregions. The microregion structure is reported as dependent on the extent of hydrogen bonding, thus, the addition of water results in structural changes and loss of volatiles. Karel (1975c) used replacement of water–carbohydrate hydrogen bonds originally present in the food by carbohydrate–carbohydrate hydrogen bonds to explain the formation of microregions. The replacement of the bonds increases the stability of the microstructures and decreases the mobility of carbohydrate molecules. This results in a reduced loss of volatiles from the microregion because of reduced permeability (Karel, 1975c).

Coumans et al. (1994) indicated that aroma retention during spray drying is lower than expected from diffusion theory because of losses prior to droplet formation caused by highly turbulent liquid on the nozzle, losses in freshly formed droplets, bad mixing conditions in the nozzles, and morphological changes in the particles. In freeze dehydration, aroma retention is better than in spray drying because of the low processing temperature. Also, Coumans et al. (1994) established the selective diffusion of aroma components homogeneously dissolved in the matrix phase and the evaporation of aroma droplets from the product matrix as the main mechanisms of aroma losses during freeze drying. These two mechanisms differ from the microregion theory proposed by Karel (1975c).

A comprehensive mathematical analysis of aroma retention in spray drying and freeze drying, based on the dependence of aroma activity on the aroma concentration (ρ_a), can be expressed as (Coumans et al., 1994):

$$\eta_a = -D_a \frac{\partial \rho_a}{\partial r} - D_a \rho_a \frac{d \ln(H_a)}{d\rho_w} \frac{\partial \rho_w}{\partial r} + w_a D_a \left(\frac{\eta_w}{D_{aw}} + \frac{\eta_s}{D_{as}} \right) \quad (100)$$

where H_a is the modified activity coefficient which depends on the water concentration $(\rho_w, \text{kg/m}^3)$, r is the distance from the product surface (m), w_a is the aroma mass fraction, w_w is the water mass fraction, η_w is the water flux (kg/m^2 s), η_s is the dissolved solid mass flux, D_{aw} is the binary diffusion coefficient for aroma–water (m^2/s), D_{as} is the binary diffusion coefficient for aroma–dissolved solids, and the aroma diffusion coefficient D_a is expressed as

$$\frac{1}{D_a} = \frac{w_w}{D_{aw}} + \frac{1-w_w}{D_{as}} \tag{101}$$

The aroma flux depends on gradients of aroma concentration and water concentration. Assuming that aroma losses occur only during the constant rate period with no shrinking of the material, mass transfer of aroma components can be described by a constant binary diffusion coefficient, and the aroma diffusion coefficient becomes zero when a critical value in water concentration is reached. The aroma retention (AR) can be expressed as follows (Coumans et al., 1994):

$$AR = 1 - 2(v+1)\left[\frac{F_{oc}}{\pi^2}\right]^{0.5} \qquad F_{oc} < 0.22 - 0.1v \tag{102}$$

$$AR = \frac{8-v}{\pi^2} \exp\left[\frac{\pi^2 F_{oc}}{4-1.5v}\right] \qquad F_{oc} \geq 0.22 - 0.1v \tag{103}$$

$$F_{oc} = D_{a,\,eff}\phi_{red} \qquad\qquad \phi_{red} = \frac{t_c}{R_o^2} \tag{104}$$

$$D_{a,\,eff} = D_{ao} \exp\left(-\frac{E_a}{RT} + f_a \frac{\rho_{wo}}{d_w}\right) \tag{105}$$

where F_{oc} is the critical Fourier number, t_c is the time to reach the critical moisture content on the product surface, R_o is the half-thickness or radius of the material, ϕ_{red} is the reduced time, $D_{a,eff}$ is the effective aroma diffusion coefficient, D_{ao}, E_a, and f_a are correlation parameters in Eq. (105), R is the gas constant, T is the absolute temperature, ρ_{wo} is the concentration of water in the food, d_w is the density of water, and v is a geometry parameter (0 for infinite extended slab, 2 for spheres).

Table 3.6.   Kinetic constants of reactions.

Nutrient	Substrate	D(min)
Thiamin (degradation)	Meat	158.0
Carotene (degradation)	Liver	43.6
Ascorbic acid (degradation)	Peas	246.0
Non-enzymatic browning	Milk	12.5
Chlorophyl (degradation)	Peas	13.2
Sensory quality loss	Various	5–500

Adapted from Toledo (1991).

Degradation of nutrients in food systems follows a first-order reaction in most cases. Table 3.6 summarizes the kinetic constants for some reactions occurring in foods. The general model is expressed as:

$$-dA/dt = kA \tag{106}$$

or
$$\ln(A/A_0) = -kt \tag{107}$$

$$\log(A/A_0) = -t/D \tag{108}$$

where A is the concentration of compound A at time t, A_0 is the initial concentration of A, k is the reaction rate constant, and D is the reciprocal of the slope of a plot of $\log(A)$ against t. D and k are related as follows:

$$D = 2.303/k \tag{109}$$

The Arrhenius equation, based on the activated complex theory for chemical reactions, relates the reaction rate constants to the absolute temperature T. The equation is expressed as:

$$k = K_0 \exp(-E_a/RT) \tag{110}$$

where K_0 is the reaction rate as T (temperature) approaches infinity, E_a is the activation energy, and R is the universal gas constant (Toledo, 1991).

Mild heating will improve the digestibility of proteins and carbohydrates. However, damages can be caused if inappropriate temperature or process times are used (Erbersdobler, 1985). The effect of drying on proteins is expressed as a decrease in the digestibility and biological value of the protein.

Lipid quality is affected by dehydration. The peroxides formed from the lipids react with proteins and vitamins, or decompose to secondary products and promote off-flavors, strong odors, and organoleptic rancidity (Karel, 1985). The exclusion of oxidizing agents is an important consideration while drying a food.

Vitamin losses are significant during drying (Erbersdobler, 1985). Thiamin, folic acid, and vitamin C are reduced in dried food. The ascorbic acid degradation varies exponentially with water activity:

$$K = b_1 \exp(b_2 a_w) \tag{111}$$

where K is the first order reaction rate constant, a_w is the water activity, and b_1 and b_2 are constants (Leung, 1987).

3.6 CONCLUDING REMARKS

The physical and chemical changes during a drying step or process will improve certain characteristics of the final product, but, in most cases, a loss of nutrients and organoleptic properties has been reported. The proper handling of those reactions and physical changes is necessary to ensure that the product has a high nutritional value as well as a significant extended shelf life. Pretreatments with combined methods prior to drying may improve textural properties and reduce degradation reactions.

The prediction of chemical and microbial stability will be a function of both water activity and the glass transition temperature. Chirife and Buera (1994) demonstrated that the water-dynamic map did not enable prediction of the microbial stability of foods and Karel et al. (1993) demonstrated the usefulness of the map to define quality modifying events such as collapse, crystallization, and stickiness during food dehydration. Thus, both concepts should be used in a complementary manner to characterize food products.

3.7 REFERENCES

Abrams, D. S. and Prausnitz, J. M. 1975. Statistical thermodynamics of liquid mixtures: a new expression for the excess Gibbs energy of partly or completely miscible systems. *AIChE J.* 21(1):116–128.

Allen, G. 1993. A history of the glassy state. In *The Glassy State in Foods*, edited by J. M. V. Blanshard and P. J. Lillford. Nottingham University Press, Loughborough, Leicestershire, UK.

Beuchat, L. R. 1983. Influence of water activity on growth, metabolic activities and survival of yeasts and molds. *J. Food Prot.* 46(2):135–141.

Bone, D. P. 1987. Practical applications of water activity and moisture relations in foods. In *Water activity: Theory and Applications to Food,* edited by L. B. Rockland and L. R. Beuchat. Marcel Dekker, New York.

Boquet, R., Chirife, J., and Iglesias, H.A. 1980. Technical note: on the equivalence of isotherm equations. *J. Food Technol.* 15:345–349.

Bromley, L. A. 1973. Thermodynamic properties of strong electrolytes in aqueous solutions. *AICHE J.* 19(2):313–320.

Bruin, S. and Luyden, K. Ch. A. M. 1980. Drying of food materials. In *Advances in Drying,* Vol. 1, edited by A. S. Mujumdar. Hemisphere Publishing, New York.

Cheftel, J. C., Cuq, J. L. and Lorient, D. 1985. Amino acids, peptides and proteins. In *Food Chemistry,* Second edition, edited by O. R. Fennema. Marcel Dekker, New York.

Chirife, J. 1987. Predicción de la actividad de agua en alimentos. In *Conservación de Alimentos de alta humedad por métodos combinados basados en la reducción de la actividad de agua.* Programa de Ciencia y Tecnología para el Desarrollo, V-Centenario CYTED-D, México.

Chirife, J. and Buera, M. P. 1994. Water activity, glass transition and microbial stability in concentrate/semimoist food systems. *J. Food Sci.* 59(5): 921–927.

Chirife, J., and Favetto, G. J. 1992. Fundamental aspects of food preservation by combined methods. Int. Union of Food Sci. and Technol.–CYTED D–Univ. de las Americas, Puebla, México.

Chirife, J., Ferro-Fontán, C., and Benmergui, E. A. 1980. The prediction of water activity in aqueous solutions in connection with intermediate moisture foods. IV. A_w prediction in aqueous non-electrolyte solutions. *J. Food. Technol.* 15:59–70.

Christensen, C. Sander, B., Fredenslund, A., and Basmussen, P. 1983. Towards the extention of UNIFAC to mixtures with electrolytes. *Fluid Phase Equilibria* 13:297–309.

Chung, D. S. and Pfost, H. B. 1967a. Adsorption and desorption of water vapor by cereal grains and their products. I. Heat and free energy changes of adsorption and desorption. *Trans. ASAE* 10:549–551.

Chung, D. S. and Pfost, H. B. 1967b. Adsorption and desorption of water vapor by cereal grains and their products. II. Development of the general isotherm equation. *Trans. ASAE* 10:552–555.

Coumans, W. J., Kerkhof, Piet J. A. M., and Bruin, S. 1994. Theoretical and practical aspects of aroma retention in spray drying and freeze-drying. *Drying Technol.* 12:99–149.

De Gois, V. A. and Cal-Vidal, J. 1986. Water sorption characteristics of freeze-dried papaya in powdered and granular forms. *Can. Inst. Food Sci. Technol. J.* 19(1):7–11.

de Man, J. M. 1982. *Principles of Food Chemistry.* AVI Publishing, Westport, CT.

Erbersdobler, H. F. 1985. Loss of nutritive value on drying. In *Concentration and Drying of Foods,* edited by D. MacCarthy. Elsevier Applied Science Publishers, New York.

Ferro-Fontán, C., Benmergui, E. A., and Chirife, J. 1980. The prediction of water activity in aqueous solutions in connection with intermediate moisture foods. III. A_w prediction in multicomponent strong electrolyte aqueous solutions. *J. Food. Technol.* 15:47–58.

Ferro-Fontán, C. and Chirife, J. 1981. Technical Note: a refinement of Ross's equation for predicting the water activity of non-electrolyte mixtures. *J. Food. Technol.* 16:219–221.

Ferro-Fontán, C., Chirife, J., and Boquet, R. 1981. Water activity in multicomponent non-electrolyte solutions. *J. Food. Technol.* 18:553–559.

FPI, 1982. Alimentos Enlatados. Principios del Control del Procesamiento Térmico, Acidificación y Evaluación del Cierre de los Envases, Second edition. Spanish version by N. Diáz and J. R. Cruz Cay. The Food Processor Institute, Washington, D.C.

Fredenslund, A., Jones, R. L., and Prausnitz, J. M. 1975. Group-contribution estimation of activity coefficients in nonideal liquid mixtures. *AIChE J.* 21(6):1086–1099.

Henderson, S. M. 1952. A basic concept of equilibrium moisture. *Agric. Eng.* 33:29–32.

Iglesias, H. A. and Chirife, J. 1978. An empirical equation for fitting water sorption isotherms of fruit and related products. *Can. Inst. Food Sci. Technol. J.* 11(1):12–15.

Jowitt, R., Escher, F., Hallstrom, B., Meffert, H. F. Th., Spiess, W., and Vos, G. 1981. *Physical Properties of Foods.* Applied Science Publishers, London, UK.

Kapsalis, J. G. 1987. Influences of hysteresis and temperature on moisture sorption isotherms. In *Water Activity: Theory and Applications to Food*, edited by L. B. Rockland and L. R. Beuchat. Marcel Dekker, New York.

Karel, M. 1975a. Water activity and food preservation. In *Principle of Food Science. Part II: Physical Principles of Food Preservation*, edited by M. Karel, O. R. Fennema, and D. B. Lund. Marcel Dekker, New York.

Karel, M. 1975b. Freeze dehydration of foods. In *Principle of Food Science. Part II: Physical Principles of Food Preservation*, edited by M. Karel, O. R. Fennema, and D. B. Lund. Marcel Dekker, New York.

Karel, M. 1975c. Dehydration of foods. In *Principle of Food Science. Part II: Physical Principles of Food Preservation*, edited by M. Karel, O. R. Fennema, and D. B. Lund. Marcel Dekker, New York.

Karel, M. 1985. Control of lipid oxidation in dried foods. In *Concentration and Drying of Foods*, edited by D. MacCarthy. Elsevier Applied Science Publishers, New York.

Karel, M., Buera, M. P., and Roos, Y. 1993. Effect of glass transition on processing and storage. In *The Glassy State in Foods*, edited by J. M. V. Blanshard and P. J. Lillford. Nottingham University Press, Loughborough, Leicestershire, UK.

Kitic, D., Pereira, D. C., Favetto, G., Resnik, S., and Chirife, J. 1986. Theoretical prediction of the water activity of standard saturated salt solutions at various temperatures. *J. Food Sci.* 51(4):1037–1040.

Kumar, M. 1974. Water vapor adsorption on whole corn flour, degermed corn flour, and germ flour. *J. Food Technol.* 9:433–444.

Labuza, T. P. 1968. Sorption phenomena in foods. *Food Technol.* 22(3): 263–266.

Labuza, T. P., McNally, L., Gallaghe, D., Hawkes, J., and Hurtado, F. 1972. Stability of intermediate moisture foods. 1. Lipid Oxidation. *J. Food Sci.* 37:154–159.

Lang, K. W. and Steinberg, M. P. 1981. Predicting water activity from 0.30 to 0.95 of a multicomponent food formulation. *J. Food Sci.* 46:670–672, 680.

Le Maguer, M. 1987. Mechanics and influence of water binding on water activity. In *Water Activity: Theory and Applications to Food*, edited by L. B. Rockland and L. R. Beuchat. Marcel Dekker, New York.

Leung, H. K. 1986. Water activity and other colligative properties of foods. In *Physical and Chemical Properties of Foods*, edited by M. R. Okos. American Society of Agricultural Engineers, St. Joseph, MI.

Leung, H. K. 1987. Influence of water activity on chemical reactivity. In *Water Activity: Theory and Applications to Food*, edited by L. B. Rockland and L. R. Beuchat. Marcel Dekker, New York.

Lima, A. W. O. and Cal-Vidal, J. 1983. Hygroscopic behavior of freeze dried bananas. *J. Food Technol.* 18:687–696.

Lindsay, R. C. 1985. Food Additives. In *Food Chemistry*, Second edition, edited by O. R. Fennema. Marcel Dekker, New York.

Money, R. W. and Born, R. 1951. Equilibrium humidity of sugar solutions. *J. Sci. Food Agric.* 2:180–185.

Monsalve-González, A., Barbosa-Cánovas, G., and Cavalieri, R. P. 1993a. Mass transfer and textural changes during processing of apples by combined methods. *J. Food Sci.* 58(5):1118–1124

Monsalve-González, A., Barbosa-Cánovas, G. V., Cavalieri, R. P., McEvily, A. J., and Iyengan, R. 1993b. Control of browning during storage of apple slices preserved by combined methods. 4-Hexylresorcinol as anti-browning agent. *J. Food Sci.* 58(4):797–800, 826.

Nawar, W. W. 1985. Lipids. In *Food Chemistry*, Second edition, edited by O. R. Fennema. Marcel Dekker, New York.

Norrish, R. S. 1966. An equation for the activity coefficients and equilibrium relative humidities of water in confectionery syrups. *J. Food. Technol.* 1:25–39.

Okos, M. R., Narsimhan, G., Singh, R. K., and Weitnauer, A. C. 1992. Food dehydration. In *Handbook of Food Engineering*, edited by D. R. Heldman and D. B. Lund. Marcel Dekker, New York.

Pitzer, K. S. 1973. Thermodynamics of electrolytes. I. Theoretical basis and general equations. *J. Phys. Chem.* 77(2):268–277.

Pitzer, K. S. 1979. Theory: ion interaction approach. In *Activity Coefficients in Electrolyte Solutions*, Vol. I, edited by R. M. Pytkowicz. CRC Press, Boca Ratón, FL.

Pitzer, K. S. and Kim, J. J. 1974. Thermodynamics of electrolytes. IV. Activity and osmotic coefficients for mixed electrolytes. *J. Am. Chem. Soc.* 96:5701–5707.

Pitzer, K. S. and Mayorga, G. 1973. Thermodynamics of electrolytes. II. Activity and osmotic coefficients for strong electrolytes with one or both ions univalent. *J. Phys. Chem.* 77(19):2300–2308.

Rockland, L.B. 1960. Saturated salt solutions for static control of relative humidity between 5 and 40 °C. *Anal. Chem.* 32(9):1375–1377.

Rockland, L. B. 1969. Water activity and storage stability. *Food Technol.* 23(10):11–17.

Rockland, L. B. and Nishi, S. K. 1980. Influence of water activity on food product quality and stability. *Food Technol.* 34(4):42–51.

Román, G. N., Urbicain, M. J., and Rotstein, E. 1982. Moisture equilibrium in apples at several temperatures: experimental data and theoretical considerations. *J. Food Sci.* 47:1484–1488.

Roos, Y. H. 1992. Phase transition and transformations in food systems. In *Handbook of Food Engineering*, edited by D. R. Heldman and D. B. Lund. Marcel Dekker, New York.

Salwin, H. and Slawson, V. 1959. Moisture transfer in combinations of dehydrated foods. *Food Technol.* 13:715–718.

Scorza, Q. C., Chirife, J., Cattaneo, P., Vigo, M. S., Bertoni, M. H., and Sarraih, P. 1981. Factores que condicionan el crecimiento microbiano en alimentos de humedad intermedia. *La alimentación Latinoamericana* 127:62–67.

Slade, L. and Levine, H. 1991. Beyond water activity: recent advances based on an alternative approach to the assesment of food quality and safety. *Crit. Rev. Food Sci. Nutr.* 30(2,3):115–360.

Slade, L. and Levine, H. 1993. Glass transitions and water–food structure interactions. Personal communication.

Smith, P. R. 1947. The sorption of water vapor by high polymers. *J. Am. Chem. Soc.* 69:646–651.

Stokes, R. H. 1979. Thermodynamics of solutions. In *Activity Coefficients in Electrolyte Solutions*, Vol. I, edited by R. M. Pytkowicz. CRC Press, Boca Ratón, FL.

Tant, M. R. and Wilkes, G. L. 1981. An overview of the nonequilibrium behavior of polymer glasses. *Polym. Eng. Sci.* 21:874–895.

Teng, T. T. and Seow, C. C. 1981. A comparative study of methods for prediction of water activity of multicomponent aqueous solutions. *J. Food. Technol.* 16:409–419.

Toledo, R. T. 1991. *Fundamentals of Food Process Engineering*, Second edition. Van Nostrand Reinhold, New York.

Troller, J. A. 1987. Adaptation and growth of microorganisms in environments with reduced water activity. In *Water Activity: Theory and Applications to Food*, edited by L. B. Rockland and L. R. Beuchat. Marcel Dekker, New York.

Troller, J. A. and Christian, J. H. B. 1978. *Water Activity in Food*. Academic Press, New York.

Van den Berg, C. 1985. Water activity. In *Concentration and Drying of Foods*, edited by D. MacCarthy. Elsevier Applied Science Publishers, New York.

Van den Berg, C. and Bruin, S. 1981. Water activity and its estimation in food systems: theoretical aspects. In *Water Activity: Influences on food quality*, edited by L. B. Rockland and G. F. Steward. Academic Press, New York.

Van Ness, H. C. and Abbott, M. M. 1982. *Classical thermodynamics of non-electrolyte solutions with applications to phase equilibria*. McGraw-Hill Chemical Eng. Series. McGraw-Hill, New York.

Vega-Mercado, H. and Barbosa-Cánovas, G. V. 1993a. Comparison of moisture sorption isotherm models in freeze-dried pineapple pulp. *J. Agric. Univ. Puerto Rico* 77(3-4):113–128.

Vega-Mercado, H. and Barbosa-Cánovas, G. V. 1993b. Heat of sorption and free energy change of freeze-dried pineapple pulp. *J. Agric. Univ. Puerto Rico* 77(3-4):101–112.

Villar, T. and Silvera, C. 1987. *Alimentos de Humedad Intermedia*. Universidad de la República, Facultad de Química, Montevideo, Uruguay.

Whistler, R. L. and Daniel, J. R. 1985. Carbohydrates. In *Food Chemistry*, Second edition, edited by O.R. Fennema. Marcel Dekker, New York.

Williams, M. L., Landel, R. F., and Ferry, J. D. 1955. The temperature dependence of relaxation mechanisms in amorphous polymers and other glass forming liquids. *J. Am. Chem. Soc.* 77:3701–3707.

Wolf, W., Spiess, W., Jung, G., Weisser, H., Bizot, H., and Duckworth, R.B. 1984. The water vapour sorption isotherms of microcrystalline cellulose (MCC) and of purified potato starch. Results of a collaborative study. *J. Food Eng.* 3:51–73.

CHAPTER 4

DEHYDRATION MECHANISMS

4.0 INTRODUCTION

This chapter discusses the simultaneous heat and mass transfer mechanisms and the moisture migration theories associated with the drying process. Drying is defined as the removal of moisture from a product, and in most practical situations the principal step during drying is the internal mass transfer (Chirife, 1983). Van Arsdel and Copley (1963) pointed out the necessity of understanding the real mechanism of the drying process for proper equipment design and best quality food products.

The mechanisms of moisture movement within the product can be summarized as (Van Arsdel and Copley, 1963): water movement under capillary forces; diffusion of liquid due to concentration gradients; surface diffusion; water vapor diffusion in air-filled pores; flow due to pressure gradient; and flow due to vaporization–condensation sequence. Capillary forces are responsible for water retention in porous solids of rigid construction, whereas osmotic pressure is responsible in aggregates of fine powders and on the surface of the solid (Toei, 1983). The vapor pressure of capillary and osmotic water is equal to the vapor pressure of pure water whereas adsorbed water has a vapor pressure lower than that of pure water. The nature of the material to be dried is a factor to consider when designing a drying process. Its physical and chemical properties play an important role during drying because of possible changes that will occur and the effect of these changes on the removal of water from the product. An hygroscopic material is one that contains bound water which exerts a vapor pressure less than that of liquid water at the same temperature. In carbohydrate-based products, hygroscopicity is expected because of the hydroxyl groups around the sugar molecules which allow hydrogen bonds with water (Whistler and Daniels, 1985). The interaction between water molecules and the hydroxyl groups leads to the solvating or solubilization of sugars. Table 4. 1 summarizes the hygroscopicity capacity of some sugars when they are exposed to moist air. The hygroscopicity capacity of sugar is also referred to as humectancy.

In water-soluble proteins such as most globular proteins the polar amino acids are uniformly distributed at the surface (Cheftel et al., 1985) whereas the hydrophobic

Table 4. 1.   Hygroscopicity of sugar (as discussed by Whistler and Daniels, 1985).

Sugar	60% RH—1 hr.	Water % absorbed at 20°C 60% RH—9 days	100% RH—25 days
D-Glucose	0.07	0.07	14.5
D-Fructose	0.28	0.63	73.4
Sucrose	0.04	0.03	18.4
Maltose	0.80	7.00	18.4
Lactose	0.54	1.20	1.40

groups tend to locate toward the inside of the molecule. This arrangement promotes the formation of hydrogen bonds with water, which explains the solubility of these types of proteins.

4.1 THE DRYING PROCESS

Drying data are usually expressed as total weight of the wet material as a function of time during the drying process, as shown in Figure 4.1. The data can be expressed in terms of rate of drying by recalculation of some values.

The moisture content is defined as the ratio of the amount of water in the food to the amount of dry solids, and is expressed as:

$$X_t = (W_t - F_s)/F_s \tag{1}$$

where W_t is the total weight of the wet material at time t, F_s is the weight of the dry solids, and X_t is the moisture expressed as weight of water/weight of dry solids. An additional quantity that is important when designing a drying process is the free moisture content, X. The free moisture content can be evaluated by considering the equilibrium moisture content X_{eq}:

$$X = X_t - X_{eq} \tag{2}$$

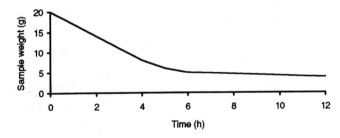

Figure 4.1. Weight change during a drying process.

A plot of X versus t is presented in Figure 4.2. The rate of drying, R, can be expressed proportional to the change in moisture content as a function of time (t):

$$R \propto dX/dt \tag{3}$$

Considering the curve presented in Figure 4.2, individual values of dX/dt as a function of time can be obtained from tangent lines drawn to the curve of X versus t. Replacing the proportionality condition in Eq. 3 by F_s/A, the drying rate can be expressed as (Geankoplis, 1983):

$$R = -(F_s/A)(dX/dt) \tag{4}$$

where R is the drying rate and A is the surface area where the drying takes place. Then a plot of R versus t gives a curve similar to that shown in Figure 4.1.

The drying process of a material can be described as a series of steps in which the drying rate plays a key role. Figure 4.3 shows a typical drying rate curve for a constant drying condition. Points A' and A represent either a hot or a cold material, respectively. Point B represents an equilibrium temperature condition of the product sur-

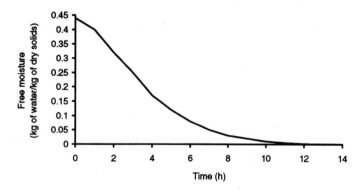

Figure 4.2. Moisture content as a function of drying time.

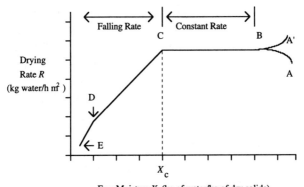

Figure 4.3.   Drying rate curve. (Adapted from Geankoplis, 1983.)

face. The period between points A (or A') and B is usually short, and is often ignored in the analysis of drying times.

Section B to C of the curve, known as the constant rate period, represents the removal of unbound water from the product. The water acts as if the solid is not present. The surface of the product is very wet at the beginning and the water activity is approximately one. On porous solids, the removed water is supplied from the interior of the solid. The constant rate period continues only as long as the water is supplied to the surface as fast as it evaporates. The temperature of the surface is approximately that of the wet bulb temperature (Geankoplis, 1983; Okos et al., 1992). In general, the drying rate is determined by external conditions of temperature, humidity, and air velocity (Chen and Johnson, 1969).

The falling rate period is reached when the drying rate starts to decrease, and the surface water activity falls to less than one. The rate of drying is governed by the internal flow of liquid or vapor (Chen and Johnson, 1969). This point is represented by C in Figure 4.3. At this point there is not enough water on the surface to maintain a water activity value of one. The falling rate period can be divided into two steps. A first falling drying rate occurs when wetted spots in the surface continually diminish until the surface is dried (point D), and a second falling rate period begins at point D when the surface is completely dry. The plane of evaporation recedes from the surface. Heat required for moisture removal is transferred through the solid to the vaporization interphase and the vapor moves through the solid into the air stream. Sometimes, there are no sharp differences between the first and second falling rate periods (Geankoplis, 1983). The amount of water removed in this period may be relatively small, while the time required may be long because the drying rate is low.

The experimental determination of rate of drying is based on a simple principle: measurement of the change in moisture content during drying. The material to be dried is placed on a tray and exposed to an air stream. The tray is suspended from a balance on a cabinet or duct through which the air flows. The weight of the material is monitored as a function of drying time. Figure 4.4 shows a typical scheme used to determine the rate of drying.

The following precautions should be observed while running a batch drying experiment:

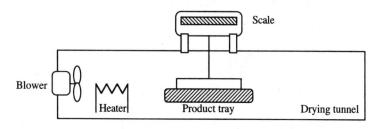

Figure 4.4. Rate of drying determination.

- The sample should not be too small.
- The drying tray should be similar to the intended one in a regular drying operation.
- Temperature, velocity, humidity, and direction of the air should be constant.

4.1.1 Constant Rate Period

The transport phenomenon during the drying of a material in the constant drying period occurs by mass transfer of water vapor from the surface of the material through an air film to the environment and heat transfer through the solids of the food. The surface remains saturated while the drying is occurring because the rate of moisture movement within the solid is sufficient. Assuming only heat transfer to the solid surface by convection from the hot air and mass transfer from the surface to the hot gas, as shown in Figure 4.5, both mass and heat transfer can be expressed as (Geankoplis, 1983):

$$q = hA(T - T_w) \qquad (5)$$

$$N_a = K_y(Y_w - Y) \qquad (6)$$

where h is the heat transfer coefficient, A is the exposed drying area, T_w is the wet bulb temperature, T is the dry-

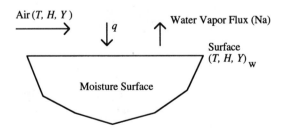

Figure 4.5.   Heat and mass transfer during drying. (Adapted from Geankoplis, 1983.)

ing temperature, N_a is the flux of water vapor, Y_w is the moisture content of the air at the surface of the solid, Y is the moisture content of the bulk air, and K_y is the mass transfer coefficient.

The heat needed to vaporize the water in the surface can be expressed as:

$$q = N_a M_a \lambda_w A \qquad (7)$$

where M_a is the molecular weight of water, λ_w is the latent heat at T_w, and the rate of drying at the constant rate period is expressed as (Okos et al., 1992):

$$R_c = k_y M_b (H_w - H) \qquad (8)$$

or

$$R_c = h(T - T_w)/\lambda_w = q/\lambda_w A \qquad (9)$$

where M_b is the molecular weight of air, H_w is the humidity at wet bulb temperature, and H is the bulk air humidity. In the absence of heat transfer by radiation or conduction, the temperature of the solid is at the wet bulb temperature of the air during the constant rate of drying period.

The mass transfer coefficient (K_y) can be obtained from the following correlation for laminar flow parallel to a flat plate (Okos *et al.*, 1992):

$$\frac{K_y l}{D_{AB}} = 0.664 \, Re^{0.5} Sc^{0.333} \qquad (10)$$

where

$$Sc = \frac{\mu}{\rho D_{AB}} \qquad Re = \frac{dv\rho}{\mu} \qquad (11)$$

where l is the length of the plate in the direction of the flow, D_{AB} is the molecular diffusivity of an air–water mixture, Re is the Reynolds number, d is a characteristic length or diameter, v is the fluid velocity, ρ is the density, μ is the viscosity, and Sc is Schmidt number. The heat transfer coefficient can be expressed as (Geankoplis, 1983):

$$h = 0.0204G^{0.8} \qquad (12)$$

where G is the mass velocity of air. The heat transfer coefficient on a slab can be represented in a Nusselt type equation (Chirife, 1983):

$$Nu = \frac{hd}{k} = 2 + \alpha Re^{0.5} Pr^{0.33} \qquad (13)$$

$$Pr = \frac{C_p \mu}{k} \qquad (14)$$

where k is the thermal conductivity, α is a constant, Re is Reynolds number, Pr is Prandlt number, and C_p is the heat capacity.

Figure 4.6 presents the stages during the constant-rate drying of a wet material as discussed by Keey (1978). During stage 1 the liquid flow may occur under a hydraulic gradient. The next stage consists of the replacement of moisture by air. At that point the surface temperature approximates that of the wet bulb temperature.

EXAMPLE 1

An insulated pan 0.5 × 0.5 m and 2.5 cm deep contains a wet granular material subjected to air drying. The sides and bottom of the pan are considered insulated. An air stream flowing over the surface provides the heat for drying by convection. The air flows at 6 m/s at 65°C with a humidity of 0.020 kg of H_2O/kg of dry air. Evaluate the constant rate of drying.

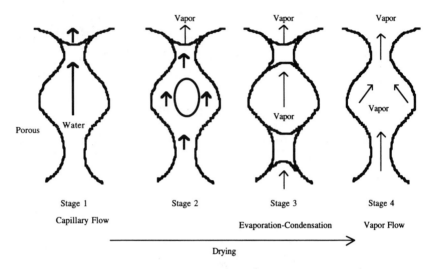

Stage 1   Stage 2   Stage 3   Stage 4

Capillary Flow   Evaporation-Condensation   Vapor Flow

Drying

Figure 4.6.   Water movement during drying of a porous material. (Adapted from Keey, 1978.)

Answer

$T = 65°C$, $H = 0.020$ kg of H_2O/kg of dry air, $T_w = 32.5°C$, and $H_w = 0.034$ kg of H_2O/kg of dry air (from the humidity chart). The humid volume is evaluated as follows:

$$v_h = (2.83E\text{-}3 + 4.56E\text{-}3H)T$$
$$= (2.83E\text{-}3 + 4.56E\text{-}3 \times 0.02)(273 + 65)$$
$$= 0.9873 \text{ m}^3/\text{kg of dry air.}$$

The density of 1 kg of air plus 0.02 kg of water:

$$\rho = (1 + 0.02)/0.9873$$
$$= 1.033 \text{ kg/m}^3.$$

The mass velocity is given by $G = v\rho$.

$$G = 6 \ (3600)(1.033)$$
$$= 22,315.40 \text{ kg/h m}^2$$

The heat transfer coefficient is evaluated from Eq. (12)

$$h = 0.204(22,315.40)^{0.8}$$
$$= 61.448 \text{W/m}^2K$$

At $T_w = 32.5$, $\lambda_w = 2423.4$ kJ/kg

Using Eq. (9):
$$R_c = h(T - T_w)/\lambda_w$$
$$= 61.45(65 - 32.5)(3600)/(2423.4E + 3)$$
$$= 2.97 \ \text{kg/m}^2\text{h}.$$

Then, the total rate of evaporation (TRE) on a surface area of 0.25 m² is:
$$\text{TRE} = R_c A$$
$$= 0.7425 \ \text{kg/h}.$$

4.1.2 Falling Rate Period

The falling rate period follows the constant rate period as stated before. The drying rate R is not constant but decreases when the moisture content passes the critical moisture content X_c. Graphical integration by plotting $1/R$ versus X will be considered to solve the integral part of Eq.(4) in the case of falling rate period.

The movement of the water within the solid may be explained by different mechanisms: diffusion of liquid due to concentration gradient, vapor diffusion due to partial vapor pressure, liquid movement due to capillary forces, effusion (Knudsen) flow, liquid movement due to gravity, and surface diffusion (Chen and Johnson, 1969; Bruin and Luyben, 1980; Fortes and Okos, 1980). The moisture movement through the food product depends on both the pore structure and interactions of moisture with the food matrix.

4.1.2.1 Diffusion Theory

The principal flow mechanism in a drying solid is diffusion as proposed elsewhere (Van Arsdel and Copley, 1963; Fortes and Okos, 1980; Geankoplis, 1983). The diffusion takes place within the solid fine structure of the solid and within the capillaries, pores, and small voids filled with vapor. This vapor diffuses outward until, at the open end of a capillary, it is carried away in the air steam. Unfortunately, the diffusion theory does not take into account shrinkage, case hardening, or sorption isotherms (Van Arsdel and Copley, 1963). The physical meaning of the diffusion coefficient is either lost or interpreted as a simultaneous effect, besides being dependent on concentration and temperature. The Fick's Law, applied to an unidimensional system shown in Figure 4.7, can be expressed as:

$$\frac{\partial X}{\partial t} = D_{eff}\frac{\partial^2 X}{\partial x^2} \tag{15}$$

where X is the free moisture content, t is the time, x is the spatial dimension, and D_{eff} is the diffusion coefficient.

The solution of the modified Fick's Law for different geometries is as follows:

A. Sphere

$$\Gamma = \frac{X - X_s}{X_0 - X_s} = \frac{6}{\pi^2}\sum_{n=1}^{\infty}\frac{1}{n^2}e^{\left(\frac{-n^2 D_{eff}t}{r^2}\right)} \tag{16}$$

where X is the moisture content at time t, X_0 is the initial moisture content, X_s is the surface moisture content, and r is the sphere radius.

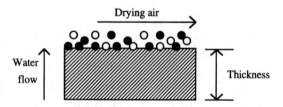

Figure 4.7.   Surface diffusion mechanisms on water vapor transport. (Adapted from Bruin and Luyben, 1980.)

B. Slab

$$\Gamma = \frac{X - X_s}{X_0 - X_s} = \frac{8}{\pi^2} \sum_{n=1}^{\infty} \frac{1}{h_n^2} e^{\left(\frac{-h_n^2 \pi^2 D_{eff} t}{4L^2}\right)}$$

$$h_n = 2n - 1 \tag{17}$$

where L is the thickness of the slab.

C. Cylinder

$$\Gamma = \frac{X - X_s}{X_0 - X_s} = \frac{4}{r_a^2} \sum_{n=1}^{\infty} \frac{1}{\beta_n^2} e^{\left(-\beta_n^2 D_{eff} t\right)} \tag{18}$$

where r_a is the radius of the cylinder and β_n is the Bessel function roots of the first kind and zero order.

The effective diffusion coefficient D_{eff} is determined by plotting experimental drying data in terms of $\ln \Gamma$ versus t. The slope of the linear segment yields the D_{eff} (Okos et al., 1992).

Two possible situations can occur within the product while it is drying: (1) liquid movement, or (2) vapor movement to the surface of the product. In the first, the water vapor flux is a function of the water concentration gradient

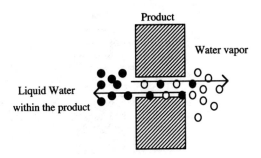

Figure 4.8. Liquid diffusion mechanisms of water vapor transport. (Adapted from Bruin and Luyben, 1980.)

within the product. The liquid water reaches the surface and is evaporated at that point, as shown in Figure 4.8.

The water vapor movement within the product may be explained in terms of the Knudsen diffusion mechanism as presented in Figure 4.9. The water flux is a function of the vapor density and the vapor diffusivity within the product. The size and amount of pores, tortuosity, and the geometry of the solid matrix affect the vapor flux as presented in the following mathematical expressions:

$$\frac{dX}{dt} = -\epsilon\tau\beta\rho D_k \tag{19}$$

$$D_k = \frac{2d}{3}\left(\frac{2RT}{\pi M}\right)^{0.5} \tag{20}$$

where ϵ is the porosity, τ is the tortuosity, β is a geometry factor, ρ is the vapor density, D_k is the Knudsen diffusivity, R is the universal gas constant, T is the absolute temperature, M is the water molecular weight, and d is the pore diameter.

The relationship between diffusivity and moisture is presented in Figure 4.10. Region A to B represents the

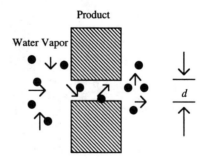

Figure 4.9.   Knudsen diffusion mechanisms of water vapor transport. (Adapted from Bruin and Luyben, 1980.)

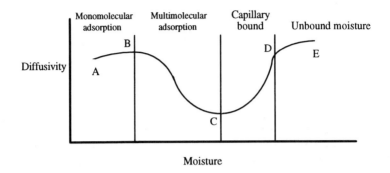

Figure 4.10.   Relationship between diffusivity and moisture content. (Adapted from Keey, 1972.)

monomolecular adsorption in the surface of the product which consists of moisture movement by vapor-phase diffusion. Region B to C covers the multimolecular adsorption where moisture begins to travel in the liquid phase. Microcapillaries play their role in region C to D, where moisture migrates easily as the finest pores are filled. Finally, in region D to E the moisture exerts its full vapor pressure and the migration of moisture is due primarily to capillarity (Keey, 1978).

The temperature dependence of the effective diffusion coefficient, D_{eff}, may be described by the Arrhenius equation:

$$D_{eff} = D_0 \exp(-E_a/R_g T) \tag{21}$$

where E_a is the activation energy, T is the absolute temperature, D_0 is a reference diffusion coefficient, and R_g is the universal gas constant. The activation energy can be determined by plotting ln $\mathbf{D_{eff}}$ versus $1/T$. Okos et al. (1992) has listed E_a values found in the literature for a variety of food products, and a summary is presented in Table 4. 2. The diffusion coefficient can be expressed as a function of moisture (Okos et al., 1992):

$$D_{\text{eff}} = D_0 \exp\left(-E_a/R_g T\right) \frac{K_{12} \exp\left(-E_b/R_g T\right)}{1 + K_{12} \exp\left(-E_b/R_g T\right)} \qquad (22)$$

$$K_{12} = k_1/k_2 \qquad (23)$$

where k_1 and k_2 are the diffusion constants at moisture levels 1 and 2 respectively, $-E_b$ is the difference in activation energy between moisture levels 1 and 2 or the binding energy, and E_a is the activation energy at the high moisture level.

Fortes and Okos (1980) stated that diffusion equations could be applied to the drying of:

Table 4. 2.   Effective diffusivity for food products.

Food	T(°C)	D_{eff} (m²/s)
Apple	66	6.40E-9
Freeze-dried apples	25	2.43E-10
Raisins	25	4.17E-11
Carrot cubes	40	6.75E-11
	60	12.1E-11
	80	17.9E-11
	100	24.1E-11
Potato	54	2.58E-11
	60	3.94E-11
	65.5	4.37E-11
	68.8	6.36E-11
Whole milk foam	50	2.0E-9
	40	1.4E-9
	35	8.5E-10
Freeze-dried ground beef	25	3.07E-11
Pears—slabs	66	9.63E-10

From Okos et al., (1992).

- Clays, starches, flours, textiles, paper, and wood that are at an equilibrium moisture content below the point of atmospheric saturation.
- Soaps, glues, and pastes (single-phased solids) in which water and solid are mutually soluble.

EXAMPLE 2

A porous solid with a critical moisture content of 0.20 kg of water/kg of dry solids is dried from 0.20 to 0.15 kg of water/kg of dry solids in 3 h. The thickness of the solid is 5 cm and the drying takes place only on one face of the solid. Calculate the drying time required for a similar product (similar moisture content range), but one that is 7 cm thick and in which the mass transfer occurs from two opposite side faces simultaneously.

Answer

A. The effective diffusion coefficient, D_{eff}, is evalauted using the data initially provided for the 5 cm thick sample. The first terms of the series from Eq. (16), a solution of Fick's Law for an unidimensional slab, are also considered. An assumption in this example is that X_s is equal to zero; all water that reaches the surface is immediately evaporated.

$$\Gamma = \frac{X}{X_o} = \frac{8}{\pi^2} e^{\left(\frac{-h_n^2 \pi^2 D_{eff} t}{4L^2} \right)}$$

where $X = 0.15$, $X_o = X_c = 0.20$, $L = 0.05$ m, $t = 10,800$ s, and $h_n = 1$. Replacing the values and solving for D_{eff}:

$$D_{eff} = 7.09E\text{-}7 \text{ m}^2/\text{s}.$$

B. The D_{eff} remains the same for the new sample, but the L value changes to 3.5 cm or half the thickness because the mass transfer occurs simultaneously through two oposite side faces. Rearranging the modified Fick's Law expression to express drying time as a function of D_{eff}, L, and Γ, the time required to dry the slab is 5300 s or 1.47 h.

4.1.2.2 Capillary Theory

The flow of a liquid through the interstices and over the surface of a solid because of molecular attraction between the liquid and the solid is referred to as capillarity (Fortes and Okos, 1980). Capillary flow has been accepted as one of the key mechanisms of drying (Fortes and Okos, 1980; Colón and Avilés, 1993). The capillary-liquid flow can be expressed as:

$$\frac{1}{A}\frac{\partial X}{\partial t} = J_L = -K_H \nabla \Psi \tag{24}$$

where Ψ is the pressure difference between the water and the air at the water–air interphase present in a capillary and K_H is the permeability. Eq. (24) can be expressed for isothermal conditions as:

$$J_L = K_{H\rho_s} \nabla X \tag{25}$$

and the permeability expressed in terms of the pore size distribution within the product:

$$K_H = (\sigma \cos \theta)/(4 \ r^2 f(r)\eta)\int r^2 f(r) \ dr \tag{26}$$

where σ is the surface tension, θ is the contact angle, $f(r)$ is a differential curve for a distribution of pore sizes by radius r, the integral is over the minimum and maximum capillary radii sizes (r_1 to r_2), η is the dynamic viscosity, ρ_s is the solid density, and X is the moisture content.

4.1.2.3 Evaporation–Condensation Theory

Water vapor within the product is condensate near the surface. This assumes that the rate of condensation is equal to the rate of evaporation at the surface of the product, and allows no acumulation of water in the pores near the surface, as presented in Figure 4.11. The theory takes into account the simultaneous diffusion of heat and mass, which assumes that the pores are a continuous network of spaces in the solid (Fortes and Okos, 1980).

The mass and heat transfer phenomena are described as follows:

Mass Balance:

$$a\tau D_v \, \nabla^2 X_v = a \frac{\partial X_v}{\partial t} + \left(1 - a\right)\rho_s \frac{\partial X}{\partial t} \qquad (27)$$

where D_v is the vapor diffusion coefficient, X_v is the vapor concentration in the pores, a is the volume fraction of air in the pores, τ is the factor taking into account the tortuosity of the diffusion path, and ρ_s is the solid skeleton density.

Energy Balance:

$$a\rho_s c_s \frac{\partial T}{\partial t} = K_T \, \nabla^2 T - q_v \frac{\partial X}{\partial t} \qquad (28)$$

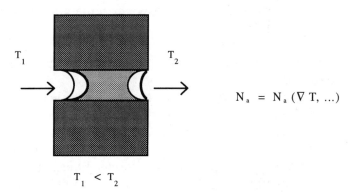

Figure 4.11.   Condensation–evaporation mechanisms of water vapor transport. (Adapted from Bruin and Luyben, 1980.)

$$\text{Le}_m = \frac{k_T\big[f + (1 - f)\rho_s\beta\big]}{D(1 - f)\rho_s\big(c_s + c_w M + \lambda_w\gamma\big)} \tag{29}$$

where c_s is the specific heat of the solid, K_T is the overall heat conduction coefficient, q_v is the heat involved in the desorption (or absorption) of water by the solid, Le_m is the modified Lewis number, k_T is the thermal conductivity, f is the void fraction of the body, β and γ are constants, D is the diffusion coefficient, c_w is the specific heat of water, and λ_w is the latent heat of vaporization. The modified Lewis number must be used to determine whether the heat transfer equation in drying problems should be considered (Young, 1969). For a modified Lewis number greater than 60, the mass transfer equation is sufficient to describe the drying process. Below this number, heat transfer becomes critical (Young, 1969).

4.1.2.4 The Luikov Theory

The phenomenon of moisture thermal diffusion discovery is credited to Luikov in 1934 (Fortes and Okos, 1980). Luikov's equations are derived based on the following assumptions: vapor, air, and water molecular and molar transport proceed simultaneously; shrinkage, contraction, and deformation are not considered; a direct relation with sorption isotherms is not considered; and isotropy is assumed.

Luikov (1975) employed the principles of irreversible thermodynamics in the development of his theory. The irreversible thermodynamics theory accounts for cross-effects between different driving forces. Fick's Law is based on the proportionality between mass transfer and the concentration gradient, without taking into account the temperature gradient. Fourier's Law is based on the proportionality between the heat transfer and the temperature gradient, without considering the concentration gradient. Thus, when more than one driving force is present, cross-effects can occur and the application of irreversible thermodynamics enables the combination of the heat and mass transfer processes.

Luikov (1966, 1975) proposed that:

$$\sigma = d_iS/dt = \Sigma J_i \cdot X_i \geq 0 \qquad (30)$$

where d_iS/dt is the rate of entropy production, J_i represents the heat and diffusion fluxes, and X_i represents the thermodynamic forces giving rise to these fluxes (temperature gradient, concentration gradient, etc.). A linear relationship between the driving forces and fluxes is expressed as (Luikov, 1966, 1975; Fortes and Okos, 1980):

$$J_i = \Sigma L_{ik} X_i \qquad (k = 1 \text{ to } n) \qquad (31)$$

where L_{ik} is a phenomenological coefficient as defined by Luikov. This equation restricts the domain of validity of Eq. (30) where the L_{ik} values are assumed constant. When the condition of linearity is not satisfied, such as in high-rate un-steady-state processes, the J_i can be expressed as:

$$\frac{1}{A}\frac{\partial X_i}{\partial t} = J_i = L_i(r)j_i + \Sigma_k\left(L_{ik}X_k + L_{ik}\frac{dX_k}{dt}\right) \qquad (32)$$

where $L_i(r)$ are constant phenomenological transfer coefficients. Considering a small rate of change on X_k, Eq. (32) can be expressed as:

$$J_i = L_i(r)J_i + \Sigma L_{ik} X_k \qquad (33)$$

In capillary porous bodies Luikov expressed the vapor and liquid fluxes as:

$$J_v = -D_v\rho_s \nabla X - K_{Tv}\rho s \nabla T \qquad (34)$$

and

$$J_l = -D_l\rho_s \nabla X - K_{Tl}\rho s \nabla T \qquad (35)$$

where D_v and D_l are the diffusion coefficients, and K_{Tv} and K_{Tl} are the thermal moisture diffusivities. By neglecting the relaxation terms in Eq. (33), the Luikov nonthermal diffusion equations for a simple case of no pressure gradient inside the body can be expressed as follows:

$$\frac{\partial X}{\partial t} = \nabla \bullet \left[D_{\mathrm{m}} \left(\nabla X + \delta \nabla T \right) \right] \tag{36}$$

$$c_b \rho_s \frac{\partial T}{\partial t} = \nabla \bullet \left(K_{\mathrm{T}} \nabla T \right) + q_v \zeta \rho_s \frac{\partial X}{\partial t} \tag{37}$$

$$D_{\mathrm{m}} = D_l + D_v \tag{38}$$

$$\delta = \frac{\left(K_{\mathrm{Tl}} + K_{\mathrm{Tv}} \right)}{D_{\mathrm{m}}} \tag{39}$$

$$q_v = h_v - h_l \tag{40}$$

where D_{m} is the total moisture diffusion coefficient, ζ is a phase conversion factor of liquid into vapor, δ is the thermal gradient coefficient, c_b is the reduced specific heat of the solid, h_v and h_l are the specific enthalpy of vapor and liquid, respectively, and q_v is the specific heat of evaporation.

In the case of intense moisture evaporation inside a capillary-porous body, there is an increase in pressure which results in a filtration pattern of vapor transfer. The equations which describe the system can be expressed as follows:

$$\frac{\partial T}{\partial t} = K_{11} \nabla^2 T + K_{12} \nabla^2 X + K_{13} \nabla^2 P \tag{41}$$

$$\frac{\partial X}{\partial t} = K_{21} \nabla^2 T + K_{22} \nabla^2 X + K_{23} \nabla^2 P \tag{42}$$

$$\frac{\partial P}{\partial t} = K_{31} \, \nabla^2 T + K_{32} \, \nabla^2 X + K_{33} \, \nabla^2 P \tag{43}$$

$$K_{11} = K_T \rho c_b + \frac{\zeta D_m \delta}{c_b} q_v \tag{44}$$

$$K_{12} = \frac{\zeta D_m}{c_b} q_v \tag{45}$$

$$K_{13} = \frac{\zeta D_m \delta_p}{c_b} q_v \tag{46}$$

$$K_{21} = D_m \delta \qquad K_{22} = D_m \qquad K_{23} = D_m \delta_p \tag{47}$$

$$K_{31} = D_m \delta \left(\frac{P\zeta}{bT} q_v + \beta - \frac{\zeta}{C_B} \right) \tag{48}$$

$$K_{32} = D_m \left(\frac{P\zeta}{C_b T} q_v + \beta - \frac{\zeta}{C_B} \right) \tag{49}$$

$$K_{33} = D_p + D_m \delta_p \left(\frac{P\zeta}{C_b T} q_v + \beta - \frac{\zeta}{C_B} \right) \tag{50}$$

$$C_B = \frac{M\epsilon b}{\rho_s RT} \tag{51}$$

$$D_p = \frac{K_p}{\rho_s C_B} \tag{52}$$

$$\beta = \frac{P\rho_s}{\rho_L \epsilon - \rho_s X} \tag{53}$$

$$\delta_p = \frac{K_p}{\rho_s D_m} \tag{54}$$

where b is the saturation level of pores with liquid, C_B is the specific body capacity for moisture, ϵ is the porosity, D_p is the convective diffusion coefficient, δ_p is the relative coefficient of filtration flow of vapor moisture, P is the pressure, and β is a coefficient that accounts for a change in the degree of filling pores and capillaries of the body with humid air.

4.1.2.5 The Philip and De Vries Theory

The mass and heat transfer under moisture and temperature gradients in porous materials can be described by the set of equations derived by Philip and De Vries (1957). Philip and De Vries (1957) proposed a vapor flux in the direction of the arrows (Figure 4.12) which results in condensation at 1 and evaporation at 2. This phenomenon tends to decrease the curvature of 1 and increase the curvature of 2. The capillary flow equals the rate of condensation at 1 and the rate of evaporation at 2.

The vapor flux is expressed as:

$$J_v = -\tau a D \left(\frac{P}{\left(P - P_w\right)} \right) \nabla \rho_w \tag{55}$$

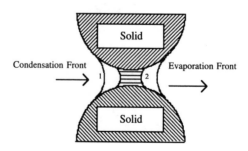

Figure 4.12.   Moisture transfer in a porous solid. (Adapted from Philip and De Vries, 1957.)

where D is the molecular diffusivity of water vapor in air, ρ_w is the water vapor density, a is the volumetric air content within the product, and τ is the tortuosity factor. The chemical potential of water in a capillary can be expressed as:

$$\Delta\mu = \Psi g \tag{56}$$

or

$$\Delta\mu = R_w T \ln\left(\frac{\rho_w}{\rho_{wo}}\right) \tag{57}$$

combining Equations (56) and (57):

$$\rho_w = \rho_{wo}a_w = \rho_{wo}e^{\left(\frac{\Psi_g}{R_w T}\right)} \tag{58}$$

Equation (58) can be transformed by considering $\rho_{wo} = \rho_{wo}$ (T), $a_w = a_w(X)$, where a_w is the water activity or relative humidity within the product, g is the gravity vector, R_w is the gas constant for water vapor, and $\partial a_w/\partial T$ is negligible:

$$\nabla\rho_w = a_w \frac{d\rho_{wo}}{dT}\nabla T + \left(\frac{g\rho_w}{R_w T}\right)\frac{\partial\Psi}{\partial X}\nabla X \tag{59}$$

At low values of H physical adsortion is dominant, while capillary condensation is the important mechanism at high values of a_w. It is reported that $a_w = 0.6$ is a transition point from physical adsortion to capillary condensation. Philip and De Vries (1957) assumed that $a_w(t)$ and $\Psi(t)$ are determined by physical adsortion for $a_w < 0.6$ and by capillarity for $a_w > 0.6$. The temperature dependence of a_w and Ψ is expressed as follows:

$$\partial a_w/\partial T \approx 0 \qquad \text{for the entire range of } a_w$$

$$\frac{\partial\Psi}{\partial T} = \left(\frac{\Psi}{\sigma}\right)\frac{\partial\sigma}{\partial T} \qquad \text{for } a_w > 0.6$$

An expression for the vapor flux in terms of concentration and thermal gradients is obtained by substituting Eq. (59) into Eq. (55) (Philip and De Vries, 1957):

$$\mathbf{J_v} = -D_v\nabla X - K_{Tv}\nabla T \tag{60}$$

where

$$K_{TV} = -\tau \mathbf{a}D\left(\frac{P}{(P-P_w)}\right)a_w\frac{d\rho_{wo}}{dT} \tag{61}$$

$$D_v = -\tau \mathbf{a}D\left(\frac{P}{P-P_w}\right)\frac{g\rho}{R_wT}\frac{\partial\Psi}{\partial X} \tag{62}$$

where K_{Tv} is the thermal vapor diffusivity, and D_v is the isothermal vapor diffusivity.

The nonisothermal liquid flux, $\mathbf{J_l}$, can be expressed as follows:

$$\mathbf{J_l} = -D_l\nabla X - K_{Tl}\nabla T - \mathbf{K_i} \tag{63}$$

$$K_{Tl} = K_H\rho_l\gamma\Psi\chi \tag{64}$$

$$D_l = K_H\rho_l(\partial\Psi/\partial X) \tag{65}$$

where K_{Tl} is the thermal liquid diffusivity, K_H is the unsaturated hydraulic conductivity, D_l is the isothermal liquid diffusivity, γ is a constant equal to $-2.9E\text{-}3°C^{-1}$, and $\mathbf{K_i}$ is the gravitational flux. By combining Eqs. (60) and (63), the total flux of water, $\mathbf{J_m}$, can be expressed as follows:

$$\mathbf{J_m} = -D_m\nabla X - K_{Tm}\nabla T - \mathbf{K_i} \tag{66}$$

$$K_{Tm} = K_{Tl} + K_{Tv} \tag{67}$$

$$D_m = D_l + D_v \tag{68}$$

where K_{Tm} is the overall thermal moisture diffusivity and D_m is the overall isothermal moisture diffusivity. A general partial differential equation that describes the moisture movement in porous materials can be expressed from Eq. (66) as:

$$\frac{\partial X}{\partial t} = \nabla \cdot \left(D_m \nabla X\right) + \nabla \cdot \left(K_{Tm} \nabla T\right) + \frac{\partial K_H}{\partial Z} \tag{69}$$

where Z is the coordinate where mass transfer occurs. The heat balance can be expressed as:

$$\rho_s c_s \frac{\partial T}{\partial t} = q_v \nabla \cdot \left(D_v \nabla X\right) + \nabla \cdot \left(K_T \nabla T\right) \tag{70}$$

The liquid diffusivities (K_{Tl}, and D_l) tend to be the most important variables at high moisture contents while the vapor difussivities (K_{TV} and D_v) are dominant at low moisture content (Philip and De Vries, 1957).

4.1.2.6 The Berger and Pei Theory

The Berger and Pei theory is based on the following assumptions (Berger and Pei, 1973): liquid transfer is due to capillary flow and concentration gradients; vapor diffusion is due to vapor pressure gradient; heat transfers through the solid skeleton by conduction; liquid content, partial vapor pressure, and temperature are in equilibrium at any location within the material; all mass and heat transfer parameters are constant; Fick's Law is valid; the vapor pressure of liquid water in excess of maximum sorptional content is equal to the saturation value; therefore, the Clausius–Clapeyron equation may be expressed as shown by Eq. (71); and the maximum sorptional liquid content is a function of temperature only as expressed by Eq. (72). The theoretical model considered by Berger and Pei (1973) is presented in Figure 4.13.

$$\rho_{vm} = \left(\frac{1}{R_w T}\right) e^{\left(58.7395 - \frac{6847.36}{T} - 5.62 \ln T\right)} \tag{71}$$

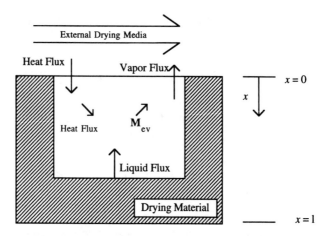

Figure 4.13.   Theoretical model discussed by Berger and Pei. (Reprinted from Berger and Pei, © 1973, p. 295, with kind permission from Elsevier Science Ltd., UK.)

$$X_{sm} = \psi_2(T) \qquad (72)$$

The liquid and vapor flux can be expressed as follows:

$$\mathbf{J}_1 = D_l \rho_l \frac{\partial X}{\partial X} \qquad (73)$$

$$\mathbf{J}_v = D_v(\epsilon - X)\frac{\partial \rho_w}{dx} \qquad (74)$$

where x is the space coordinate for heat and mass transfer, ϵ is the void fraction of the solid, D_l is the liquid conductivity, D_v is the vapor diffusivity, ρ_w is the vapor density, ρ_l is the liquid density, and X is the liquid content. Considering a mass balance over a small element of volume, the differential equation for the liquid transfer can be expressed as:

$$D_l\rho_l \frac{\partial^2 X}{\partial x^2} - M_{ev} = \rho_l \frac{\partial X}{\partial t} \qquad (75)$$

where M_{ev} is the rate of evaporation inside the drying material, and x is the thickness. The differential equation for the vapor transfer is expressed as:

$$D_v \frac{\partial \left((\epsilon - X) \frac{\partial \rho_w}{\partial x} \right)}{\partial x} + M_{ev} = \frac{\partial \left[(\epsilon - X) \rho_w \right]}{\partial t} \tag{76}$$

The total moisture transfer can be expressed as (Berger and Pei, 1973):

$$D_l \rho_l \frac{\partial^2 X}{\partial x^2} + D_v \left[(\epsilon - X) \left(\frac{\partial^2 \rho_w}{\partial x^2} \right) - \left(\frac{\partial X}{\partial x} \right) \left(\frac{\partial \rho_w}{\partial x} \right) \right]$$

$$= \left(\rho_l - \rho_w \right) \left(\frac{\partial X}{\partial t} \right) + \left(\epsilon - X \right) \left(\frac{\partial \rho_w}{\partial t} \right) \tag{77}$$

and the heat balance leads to:

$$\alpha \frac{\partial^2 T}{\partial x^2} + \frac{q_v}{\rho_s c_s} \left\{ D_v \left[(\epsilon - X) \left(\frac{\partial^2 \rho_w}{\partial x^2} \right) - \left(\frac{\partial X}{\partial x} \right) \left(\frac{\partial \rho_w}{\partial x} \right) \right] \right.$$

$$\left. + \rho_w \frac{\partial X}{\partial t} - (\epsilon - X) \frac{\partial \rho_w}{\partial t} \right\} = \frac{\partial T}{\partial t} \tag{78}$$

The solution of Eq. (77) and (78) is based on the following boundary conditions:

The boundary condition for Eq. (77) at $x = 0$ is expressed as:

$$D_l \rho_l \frac{\partial X}{\partial x} + D_v (\epsilon - X) \frac{\partial \rho_v}{\partial x} = D_c \left(\rho_v - \rho_{va} \right) \tag{79}$$

Equation (79) represents the total vapor flux away from the surface into the drying media, which is equal to the sum of liquid and vapor flux to the surface of the product. The rate of drying can be expressed as:

$$\text{Rate} = K_c(\rho_v \big|_{x=0} - \rho_{va}) \tag{80}$$

The boundary condition for Eq. (78) at $x = 0$ is expressed as:

$$h(T_a - T) = q_v D_l \rho_l \frac{\partial X}{\partial x} - K_T \frac{\partial T}{\partial x} \tag{81}$$

where $q_v D_l \rho_l \dfrac{\partial X}{\partial x}$ denotes the amount of heat required to evaporate the liquid at the surface of the product. Finally, Berger and Pei (1957) assumed that no mass and heat are transferred across that surface of the drying material and that the boundary conditions at $x = 1$ are expressed for Eqs. (77) and (78) as follows:

$$D_l \rho_l \frac{\partial X}{\partial x} = D_v(\epsilon - X)\frac{\partial \rho_v}{\partial x} = 0 \tag{82}$$

$$\frac{\partial T}{\partial x} = 0 \tag{83}$$

The final solution of the differential equation is expressed as a function of a dimensionless group defined by Berger and Pei (1973). This solution demonstrated that the drying behavior during the falling rate period is to a large extent controlled by the sorption isotherm of the product.

4.1.2.7 The Whitaker Theory

Whitaker (1980) described the transport of heat, mass, and momentum in a porous media by the *volume averaging* method. Whitaker approached the transport problem using the microscopic transport equations for continuous media. These equations were applied to the solid, liquid, and gas phases. Through the method of volume averaging, the theory used the microscopic approach in the macroscopic form. The basic assumptions made by Whitaker are that the effect of gas phase motion on the liquid flow during drying is neglected and the capillary pressure is only a function of temperature and saturation. The laws of continuum physics are expressed as follows:

$$\partial \rho_i / \partial t + \nabla \cdot (\rho_i v_i) = 0 \tag{84}$$

$$-\nabla p + \rho \mathbf{g} + \eta \nabla^2 v = 0 \tag{85}$$

$$\rho \, C_p \partial T / \partial t = \lambda \, \nabla^2 T + \Phi \tag{86}$$

where ρ is the total density, λ is the latent heat, ρ_i is the individual density, t is the time, v_i is the individual mass velocity, η is the viscosity, v is the mass average velocity, p is the pressure, $\mathbf{g}$ is the gravity vector, C_p is the heat capacity, T is the temperature, and Φ represents the source or sink of electromagnetic energy.

By considering a granular medium as presented in Figure 4. 14, Eqs. (84), (85), and (86) can be expressed for each of the phases (solid, liquid, and gas) as follows:

Solid Phase (s-phase)
Only the thermal energy equation is used in the solid phase:

$$\rho_\sigma C_{p\sigma} \frac{\partial T_\sigma}{\partial t} = \lambda_\sigma \nabla^2 T_\sigma + \Phi_\sigma \tag{87}$$

Liquid Phase (b-phase)
It is assumed that the liquid phase density is constant so that the continuity equation can be expressed as:

$$\nabla \cdot (v\beta) = 0 \tag{88}$$

and the momentum and thermal energy equations are expressed as:

$$-\nabla p\beta + \rho_\beta \mathbf{g} + \eta\beta \nabla^2 v\beta = 0 \tag{89}$$

$$\rho_\beta C_{p\beta}\left(\frac{\partial T_\beta}{\partial t}+v_\beta\,\nabla_\beta\right)=\lambda_\beta\,\nabla^2 T_\beta+\Phi_\beta \tag{90}$$

Gas Phase (γ-phase)

The overall continuity equation can be expressed as:

$$\frac{\partial\rho_\gamma}{\partial t}+\nabla\cdot\left(\rho_\gamma v_\gamma\right)=0 \tag{91}$$

while the continuity equation for the evaporating species is expressed as:

$$\frac{\partial\rho_1}{\partial t}+\nabla\cdot\left(\rho_1 v_\gamma\right)=\nabla\left(\rho_\gamma D\,\nabla\left(\frac{\rho_1}{\rho_\gamma}\right)\right) \tag{92}$$

where ρ_1 is the density of the evaporating species and D is the molecular diffusivity. The thermal-energy and momentum equations for the gas phase are similar to those for the liquid phase:

$$\rho_\gamma C_{p\gamma}\left(\frac{\partial T_\gamma}{\partial t}+v_\gamma\,\nabla T_\gamma\right)=\lambda_\gamma\,\nabla^2 T_\gamma+\Phi_\gamma \tag{93}$$

$$-\nabla p\gamma+\rho\gamma\,\mathbf{g}+\eta\gamma\,\nabla^2 v\gamma=0 \tag{94}$$

The heat and mass transport in a porous media can be expressed as follows (Whitaker, 1980):

Mass Transport

$$\frac{\partial\left(\Psi_\gamma<\rho>_\gamma\right)}{\partial t}+\nabla\cdot\left(<\rho_1>_\gamma<v_\gamma>\right)+\frac{1}{v}\int\rho_1\left(v_1-w\right)\mathbf{n}_{\gamma\beta}\,dA$$

$$=\nabla\cdot\left(<\rho_\gamma>_\gamma D_{\text{eff}}\cdot\nabla<\rho_1>_\gamma/<\rho_\gamma>_\gamma\right) \tag{95}$$

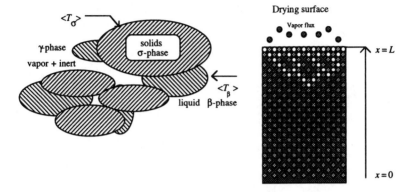

Figure 4.14. Drying of a granular porous medium. (Adapted from Whitaker, 1980).

where D_{eff} is the gas phase effective diffusivity, $< >$ means the average value, Ψ is the volume fraction of the phase, A is the area, and $\mathbf{n}$ is a unit normal vector.

Energy Transport

$$< \rho > C_p \frac{\partial < T >}{\partial t} + \left[\rho_\beta \left((C_p)_\beta < v_\beta > \right) + < \rho_\gamma >_\gamma < C_p >_\gamma < v_\gamma > \right] \nabla < T > + \Delta h_{vap} < \mathbf{m} >$$

$$= \nabla \cdot \left(\mathbf{K}_{eff}^T \nabla < T > \right) + < \Phi > \qquad (96)$$

$$< \rho > = \Psi_\sigma \rho_\sigma + \Psi_\beta \rho_\beta + \Psi_\gamma \Sigma < \rho_i >_\gamma \qquad (97)$$

$$C_p = \frac{\left[\Psi_\sigma \rho_\sigma (C_p)_\sigma + \Psi_\beta \rho_\beta (C_p)_\beta + \Psi_\gamma \Sigma < \rho_i >_\gamma (C_{pi}) \right]}{< \rho >} \qquad (98)$$

where $\mathbf{K}_{eff}^T$ is the total thermal conductivity tensor which takes into account both conduction and dispersion of thermal energy, $< \rho >$ is the spatial average density, and $\mathbf{m}$ is the mass rate of evaporation.

Momentum Transport
Gas phase

$$\langle v_\gamma \rangle = (-1/\eta\gamma)\mathbf{K}\gamma \cdot (\nabla\langle p_\gamma \rangle_\gamma - \rho_\gamma \mathbf{g}) \tag{99}$$

Liquid phase

$$\langle v_\beta \rangle = (-1/\eta_\beta)\mathbf{K}_\beta(\nabla\langle p_\beta \rangle_\beta - \rho_\beta \mathbf{g}) \tag{100}$$

and the capillary pressure ($\langle p_c \rangle$) is defined as:

$$\langle p_c \rangle = \langle p_\gamma \rangle_\gamma - \langle p_\beta \rangle_\beta \tag{101}$$

Finally, the transport equations and thermodynamic relationships for a porous body undergoing drying can be summarized as (Whitaker, 1980):

Total Thermal-Energy Equation

$$\langle \rho \rangle C_p \frac{\partial \langle T \rangle}{\partial t} - \left[\rho_\beta \left(C_p \right)_\beta \left(1 - \Psi_\sigma \right) \left(\mathbf{K_s} \cdot \nabla X + \mathbf{K}_{\langle T \rangle} \cdot \nabla \langle T \rangle - \mathbf{K_g} \cdot \mathbf{g} \right) \right.$$

$$+ \rho_\beta \left(1 - \Psi_\sigma \right) \langle \left(C_p \right)_l \rangle_\gamma \left(\mathbf{H} \frac{\partial \zeta}{\partial X} \cdot \nabla X + \mathbf{H} \frac{\partial \zeta}{\partial \langle T \rangle} \cdot \nabla \langle T \rangle \right) \right] \cdot \nabla \langle T \rangle$$

$$- \rho_\beta \left(1 - \Psi_\sigma \right) \Delta h_{\text{vap}} \nabla \cdot \left(\mathbf{H} \frac{\partial \zeta}{\partial X} \cdot \nabla X + \mathbf{H} \frac{\partial \zeta}{\partial \langle T \rangle} \cdot \nabla \langle T \rangle \right)$$

$$= \nabla \cdot \left(\mathbf{K}_{\text{eff}}^{\mathbf{T}} \cdot \nabla \langle T \rangle \right) + \langle \Phi \rangle \tag{102}$$

where $\mathbf{K_s}$ is the saturation permeability tensor, $\mathbf{K}_{\langle T \rangle}$ is the thermal permeability tensor, $\mathbf{K_g}$ is the gravitational permeability tensor, Δh_{vap} is the enthalpy of vaporization per unit mass, and X is the saturation defined as

$$\frac{\left(\rho_\beta \Psi_\beta + \langle \rho_l \rangle_\gamma \Psi_\gamma \right)}{\rho_\beta \left(\Psi_\beta + \Psi_\gamma \right)}.$$

Thermal-Energy-Flux Boundary Condition

$$[\rho\beta(1 - \Psi_\sigma)(K_s \cdot \nabla X + K_{<T>} \cdot \nabla<T> - K_g \cdot g)\Delta h_{vap} - K_{eff}^T \cdot \nabla<T>] \cdot \mathbf{n}$$

$$= \alpha (<T> - T_\infty) \tag{103}$$

where α is the heat transfer coefficient.

Total Moisture-Transport Equation

$$\frac{\partial X}{\partial t} = \nabla \cdot \left[\left(\mathbf{K_s} + \mathbf{H}\frac{\partial \zeta}{\partial X} \right) \cdot \nabla X + \left(\mathbf{K}_{<T>} + \mathbf{H}\frac{\partial \zeta}{\partial <T>} \right) \cdot \nabla <T> - \mathbf{K_g} \cdot \mathbf{g} \right] \tag{104}$$

Total Moisture-Flux Boundary Condition

$$-\rho_\beta\left(1 - \Psi_\sigma\right)\left[\left(\mathbf{K_s} + \mathbf{H}\frac{\partial \zeta}{\partial X} \right) \cdot \nabla X + \left(\mathbf{K}_{<T>} + \mathbf{H}\frac{\partial \zeta}{\partial <T>} \right) \cdot \nabla <T> - \mathbf{K_g} \cdot \mathbf{g} \right] \cdot \mathbf{n}$$

$$= \mathbf{D}\left[<\rho_1>_\gamma - \left(\rho_1\right)_\infty \right] \tag{105}$$

where **D** is the mass transfer coefficient.

Thermodynamic Relations

$$<\rho_1>_\gamma = \frac{P_1^o}{R_1 <T>} \exp\left\{ -\left[\frac{2\sigma}{\left(r\rho_\beta R_1 <T>\right)} + \frac{\Delta h_{vap}}{<T> R_1} - \frac{1}{T_o} \right] \right\} \tag{106}$$

$$<\rho_\gamma>_\gamma = <\rho_1>_\gamma \left(1 - \frac{R_1}{R_2} \right) + \frac{P_o}{<T> R_2} \tag{107}$$

where P_1^o is the partial pressure of component 1, σ is the surface tension, R_1 and R_2 are the gas constants for species 1 and 2, r is the radius of curvature, T_0 is a reference temperature, and P_0 is the ambient pressure.

The Functions H and ζ

$$\mathbf{H}(X, <T>) = \frac{<\rho_\gamma>_\gamma D_{\text{eff}}^{*1}}{\rho_\beta(1-\Psi_\sigma)\left(1 - \dfrac{<\rho_1>_\gamma}{<\rho_\gamma>}\right)} \tag{108}$$

$$\zeta(X, <T>) = \frac{<\rho_1>_\gamma}{<\rho_\gamma>_\gamma} \tag{109}$$

where D_{eff}^{*1} is the gas phase effective diffusivity tensor for diffusion of species through stagnant species 2.

A system consisting of a bed of height L and undergoing steady-state drying in one dimension can be described using the Whitaker model (Aviles, 1989). An important assumption is that the moisture level in the body is above the equilibrium moisture content. This represents a continuous liquid phase. The capillary action causes the liquid to move toward the drying surface, and thus the gas phase mass transport becomes negligible. Therefore, the governing differential equations for temperature and moisture are expressed as:

$$C_p\rho_\beta(1-\Psi_\sigma)\left[\mathbf{K_s}\frac{dX}{dx} + \mathbf{K_{<T>}}\frac{d<T>}{dx} - \mathbf{K_g}\cdot g\right]\frac{d<T>}{dx} = \frac{d}{dx}\left[\mathbf{K_{\text{eff}}^T}\cdot\frac{dT}{dx}\right] \tag{110}$$

$$\frac{d}{dx}\left[\mathbf{K_s}\frac{dX}{dx} + \mathbf{K_{<T>}}\frac{d<T>}{dx} - \mathbf{K_g}\cdot g\right] = 0 \tag{111}$$

and the liquid moisture flux can be expressed as follows:

$$<m> = -\rho_\beta(1-\Psi_\sigma)\left[\mathbf{K_s}\frac{dX}{dx} + \mathbf{K_{<T>}}\frac{d<T>}{dx} - \mathbf{K_g}\cdot g\right] \tag{112}$$

The boundary conditions at the interface between porous medium and air, $x = L$, can be expressed from Eqs. (103) and (105) as follows:

$$\rho_\beta(1 - \Psi_\sigma)\left(\mathbf{K_s} \cdot \frac{dX}{dx} + \mathbf{K_{<T>}} \cdot \frac{d<T>}{dx} - \mathbf{K_g} \cdot \mathbf{g}\right)\Delta h_{vap}\mathbf{K}_{eff}^T \cdot \frac{d}{d}$$

$$= \alpha(<T> - T_\infty) + <\Phi> \tag{113}$$

and

$$-\rho_\beta(1 - \Psi_\sigma)\left[\left(\mathbf{K_s} \frac{\partial X}{\partial x}\right) + \left(\mathbf{K_{<T>}} \frac{d<T>}{dx}\right) - \mathbf{K_g} \cdot \mathbf{g}\right]$$

$$= \mathbf{D}\left[<\rho_1>_\gamma - (\rho_1)_\infty\right] \tag{114}$$

The boundary conditions at the bottom of the packed bed, $x = 0$, are:

$$X = 1 \text{ and } T = T_o$$

The permeability tensors and moisture content are defined as follows:

$$\mathbf{K_s} = -\frac{\mathbf{K_\beta}}{\mu_\beta(1 - \Psi_\sigma)}\frac{\partial P_c}{\partial X} \tag{115}$$

$$\mathbf{K_{<T>}} = -\frac{\mathbf{K_\beta}}{\mu_\beta(1 - \Psi_\sigma)}\frac{\partial P_c}{\partial <T>} \tag{116}$$

$$\mathbf{K_g} = \frac{\rho_\beta\mathbf{K_\beta}}{\mu_\beta(1 - \Psi_\sigma)} \tag{117}$$

$$X = \frac{\rho_\beta\Psi_\beta + \rho_\gamma\Psi_\gamma}{\rho_\beta(1 - \Psi_\sigma)} \tag{118}$$

By replacing the definition of the permeability tensors and moisture content into Eqs. (110) through (114), the equations governing the transport phenomena for the temperature and moisture can be expressed in a dimensionless form as follows (Aviles, 1989):

$$U\left[H_x \frac{dX}{dX_L} + H_T \frac{dT}{dX_L} - 1\right]\frac{dT}{dX_L} = \frac{d^2T}{dX_L^2} \tag{119}$$

$$\frac{d}{dX_L}\left\{U\left[H_x \frac{dX}{dX_L} + H_T \frac{dT}{dX_L} - 1\right]\right\} = 0 \tag{120}$$

And, the boundary conditions in the bed are (Aviles, 1989):

A. At the porous medium–air interface, $X_L = 1$ or $x = L$:

$$-VU\left[H_x \frac{dX}{dX_L} + H_T \frac{dT}{dX_L} - 1\right] + \frac{dT}{dX_L} = \eta \tag{121}$$

B. At the bottom of the bed, $X_L = 0$ or $x = 0$:

$$T = T_0/T_\infty \qquad X = 1 \tag{122}$$

The steady-state drying rate is equal at any point within the bed and it is expressed as follows (Aviles, 1989):

$$U\left[H_x \frac{dX}{dX_L} + H_T \frac{dT}{dX_L} - 1\right] = \Omega \tag{123}$$

where the dimensionless variables are:

$$T = \frac{<T>}{T_\infty} \qquad\qquad H_c = \frac{h_c}{L} \tag{124}$$

$$X_L = \frac{x}{L} \qquad\qquad H_x = \frac{\partial H_c}{\partial X} \tag{125}$$

$$H_{\mathrm{T}} = \frac{\partial H_{\mathrm{c}}}{\partial T} \qquad\qquad U = \frac{\rho_\beta^2 \mathbf{K}_\beta^{\mathrm{o}} g K_r L < C_{\mathrm{p}} >_\beta}{\mu_\beta \mathbf{K}_{\mathrm{eff}}} \qquad (126)$$

$$V = \frac{\Delta h_{\mathrm{vap}}}{\mathrm{T}_\infty < C_{\mathrm{p}} >_\beta} \qquad \eta = \left[\alpha(T - 1) + \frac{I}{\mathrm{T}_\infty} \right] \frac{L}{\mathbf{K}_{\mathrm{eff}}} \qquad (127)$$

$$\Omega = \frac{D_{\mathrm{m}} \left(< \rho_\beta >_\gamma - < \rho_\beta >_\infty \right) L < C_{\mathrm{p}} >_\gamma}{\mathbf{K}_{\mathrm{eff}}} \qquad (128)$$

where h_{c} is the capillary pressure-moisture relationship which can be expressed in a nonlinear form as (Aviles, 1989):

$$h_{\mathrm{c}} = b_1 \left(1 - e^{\left(-b_2 (1-X) \right)} \right) + b_3 (1 - X) + \frac{b_4}{X^{b_5}} \qquad (129)$$

and b_i's are evaluated from experimental data, K_r is the ratio of the effective permeability ($K\beta$) for the liquid phase to the absolute permeability of the medium (K_β^{o}), or expressed in terms of the moisture content:

$$K_r = \frac{\mathbf{K}_\beta}{\mathbf{K}_\beta^{\mathrm{o}}} = \left[\frac{X - X_{\mathrm{o}}}{1 - X_{\mathrm{o}}} \right] \qquad (130)$$

and X_{o} is the equilibrium moisture content, I is the radiant intensity over the surface, D_{m} is expressed in terms of experimental liquid flux (J_{exp}) across the bed (or product):

$$D_{\mathrm{m}} = \frac{\mathbf{J}_{\mathrm{exp}}}{< \rho_\beta >_\gamma - < \rho_\beta >_\infty} \qquad (131)$$

and α is also evaluated from the experimental conditions as follows:

$$\alpha = \frac{\mathbf{J}_{\mathrm{exp}} \Delta h_{\mathrm{vap}}}{T_\infty - T_\sigma} \qquad (132)$$

Finally, an equation for the thermal energy can be obtained by combining Eqs. (119) and (128):

$$\frac{d^2T}{dX_L^2} - \Omega \frac{dT}{X_L} = 0 \qquad (133)$$

with the following solution:

$$T = C_1 + C_2 \exp(\Omega X_L) \qquad (134)$$

Applying the boundary conditions, and Eqs. (121) and (122), the temperature profile in a steady-state drying process can be expressed as:

$$T = \frac{T_o}{T_\infty} + \left(\frac{V\Omega + \eta}{\Omega}\right) \exp(-\Omega)\left[\exp(-\Omega X_L) - 1\right] \qquad (135)$$

and the moisture gradient dX/dX_L is obtained by rearranging Eq. (128):

$$\frac{dX}{dX_L} = \frac{\Omega}{UH_x}\left(1 - H_T \frac{dT}{dX_L}\right) \qquad (136)$$

where the term dT/dX_L is the derivative of Eq. (135). Since Eq. (136) is a highly nonlinear differential equation, a direct solution is hard to find. Aviles (1989) used the first three terms of a Taylor series to determine the solution of the moisture distribution equation considering a general situation $X_L = X_{Lo}$:

$$X = X_1\left|\frac{dX}{dX_L} + \frac{dX}{dX_L}\right|(X_L - X_{Lo}) + \frac{d^2X}{dX_L^2}\left|\frac{(X_L - X_{Lo})^2}{2}\right.$$

$$+ \frac{d^3X}{dX_L^3}\left|\frac{(X_L - X_{Lo})^3}{6}\right. \qquad (137)$$

Finally, Eq. (137) can be expressed in terms of the boundary condition at the bottom of the bed, $X_L = 0$:

$$X = 1 + \frac{dX}{dX_L}\bigg|(X_L) + \frac{d^2X}{dX_L^2}\bigg|\left(\frac{X_L^2}{2}\right) + \frac{d^3X}{dX_L^3}\bigg|\frac{\left(X_L^3\right)}{6} \tag{138}$$

The first, second, and third derivatives are evalauted at $X_L = 0$ using Equation (136). As shown in Eqs. (135) and (136), the thickness of the bed gives different temperature and moisture profiles. Also, the size of capillaries plays an important role as shown in Eqs. (128), (129), (135), and (136). The latter is important when considering the Whitaker model for predicting the drying process of foodstuffs.

4.1.2.8 Strongin-Borde Model

Strongin and Borde (1987) applied isotherm equations to directly determine drying-rate curves of capillary porous materials. The moisture transfer equations are expressed as:

$$\rho \, \partial X/\partial t = \rho \nabla(D \, \nabla X) - I \tag{139}$$

$$\partial(\rho_w \psi_w)/\partial t = \nabla(D_v \, \nabla \rho_w) + I \tag{140}$$

where ρ is the density of the dried material, X is the moisture content, t is time, D is the moisture diffusivity, D_v is the vapor diffusivity, ρ_w is the vapor density, ψ_w is the volume of vapor per unit volume of material, and I is the volumetric capacity of local evaporation. The main assumptions in this model are:

- Constant density and no shrinkage during drying.
- Small pressure gradient within the body.
- Constant temperature during moisture transfer.
- Liquid-vapor relationship expressed in terms of a sorption isotherm:

$$\phi_w(X) = 1 \qquad X > X_m$$

$$\phi_w(X) = \phi_{we} + k_1(X - X_e) \qquad X_m \geq X \geq X_e$$

$$\phi_w(X) = \phi_{we} - k_2(X_e - X) \qquad (X_e \geq X \geq X_d)$$

where $k_1 = (1 - \phi_{we})/((X_m - X_e)$ and $k_2 = (\phi_{we} - \phi_{wd})/(X_e - X_d)$, ϕ_w is the vapor relative density, X_m is the maximum moisture content, ϕ_{we} and X_e are the break points of approximated isotherms, k_1 and k_2 are the slopes of the isotherm straight segments, and ϕ_{wd} and X_d are the isotherm parameters at the monolayer conditions. Figure 4. 15 summarizes the above information.

- The monolayer region is not considered.
- The diffusivities, D and D_v, are equal to a mean value in each range of the isotherm.
- The vapor density and volumetric fraction are expressed as:

$$\psi_v = \psi - (\rho/\rho_l)X$$

$$\rho_v = \rho_s \phi_w$$

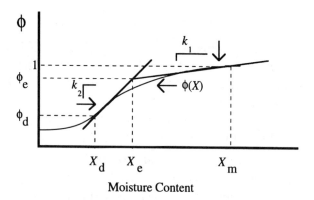

Figure 4.15. Sorption isotherm ϕ_w and its approximations. (Adapted from Strongin and Borde, 1987.)

where ρ_s is the dry solid density and ρ_l is the liquid density.

Utilizing the assumptions in Eqs. (139) and (140):

$$(1 - \rho_s/\rho_l)\rho \ \partial X/\partial t = D\rho \ \nabla^2 X \qquad (X > X_m) \qquad (141)$$

$$I = -(\rho_s/\rho_l)\rho \ \partial X/\partial t \qquad (142)$$

$$[1 + (\rho_s/\rho)k_1\psi_{v1} - (\rho_s/\rho_l)\phi_1]\rho \ \partial X_1/\partial t = D_1\rho \ \nabla^2 X_1$$
$$(X_m \geq X_1 \geq X_e) \qquad (143)$$

$$[1 + (\rho_s/\rho)k_2\psi_{v2} - (\rho_s/\rho_l)\phi_2]\rho \ \partial X_2/\partial t = D_2\rho \ \nabla^2 X_2$$
$$(X_e \geq X_2 \geq X_d) \qquad (144)$$

The major attribute of this model is the accounting of the sorption characteristics of the product by incorporating k_1 and k_2 as variables when defining the differential equations used to describe the drying process.

4.1.2.9 Regular Regime Theory

Kerkhof (1994) proposed the use of a modified Biot number to determine whether the drying process is externally controlled—the water activity at the surface decreases according to the sorption isotherm—or the process is internally controlled—the surface water concentration drops off rapidly and it is the internal rate of diffusion that determines the drying rate. The proposed Biot number is:

$$\text{Bi} = \frac{k_{g,\text{eff}} L \rho_{wg}^s}{D_{\text{eff}} \rho_s x_{cr}^*} \qquad (145)$$

where $k_{g,\text{eff}}$ is the effective mass transfer coefficient (m/s), which takes into account the change in moisture content of the air during the drying process, L is the dimension of the product particle (m), ρ_{wg}^s is the saturation vapor concentration (kg/m^3), D_{eff} is the effective diffusion coefficient (m^2/s), ρ_s is the concentration of solids (kg/m^3), and

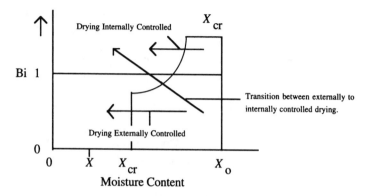

Figure 4.16.   Representation of dominating resistance as a function of the Biot number. (Reprinted from Kerkhof (1994), p. 8 by courtesy of Marcel Dekker, Inc.)

x_{cr}^* is the critical moisture content of the product. Figure 4.16 summarizes the concept proposed by Kerkhof (1994).

The moisture and energy balance for the system described in Figure 4.17 can be expressed as:

Mass:

$$F_{s\,in}X_{in} - F_{s\,out}X_{out} = \frac{d\left(M_p X_{avg}\right)}{dt} + GF_{s\,fines}X_{w\,fines} + V\left(Y_{out} - T_{in}\right) \quad (146)$$

where F_s is the solids feed (kg/h), X is the moisture content, avg is the average value, fines is the particulate carried by the air, M_p is the dry solid holdup in the dryer (kg), G is the dry air stream (kg dry air/h), and Y is the air moisture content.

Energy:

$$F_{s\,in}h_{in}^l - F_{s\,out}h_{out}^l = \frac{d\left(M_p h_{avg}^l\right)}{dt} + Vh_{fines}X_{w\,fines} + V\left(h_{out}^v h_{in}^v\right) + \frac{d\left(h_{eq}\right)}{dt} + q_{loss} \quad (147)$$

where h is the enthalpy (l is for liquid, v for vapor, eq is for equipment).

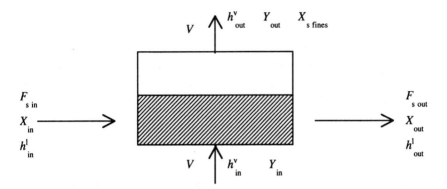

Figure 4.17. Kerkhof representation of a dryer. (Reprinted from Kerkhof (1994), p. 12 by courtesy of Marcel Dekker, Inc.)

Equations (146) and (147) can be simplified by considering the case of continuous steady-state drying, neglecting the fines carryover and heat losses:

$$F_s(X_{in} - X_{out}) = V(Y_{out} - Y_{in}) \qquad (148)$$

$$F_s(h_{in}^l - h_{out}^l) = V(h_{out}^v - h_{in}^v) \qquad (149)$$

The change in moisture content in the product can be expressed as:

$$\frac{\partial X}{\partial t} = \frac{1}{r^v}\frac{\partial}{\partial r}\left[r^v\left(D_w\frac{\partial X}{\partial r} - V^* X\right)\right] \qquad (150)$$

where X is the water concentration (kg/cm^3), t is the time, r is the thickness, v is a geometry factor (0 = slab, 1 = cylinder, 2 = sphere), D_w is the effective diffusion coefficient of water and solids, and V^* is the volume-average velocity.

At the external interface, there is an equality between the internal and external water flux that can be expressed as:

$$J_{w,i} = -D_w \nabla X + (v^* - v^i)X = k_{eff}(Y_i - Y_b) \qquad (151)$$

where K_{eff} is the effective mass transfer coefficient, Y_i is the water content of air at the interface, Y_b is the water content of bulk air, v^i is the velocity of the interface. Considering an equilibrium condition between phases:

$$Y_i = \mathbf{A}_{wi} Y_{i\,sat} = \mathbf{A}_{wi}\, \frac{P_w^{sat} M_w}{RT} \qquad (152)$$

where P_w^{sat} is the water vapor saturation pressure (Pa), Y_{isat} is the water concentration in air at a saturation condition, M_w is the water molecular weight, R is the gas constant, T is the temperature, and a_w is the water activity.

A convenient way of presenting the drying curve is defining a flux parameter as a function of moisture content (Kerkhof, 1994):

$$\mathbf{F} = J_w R_s \qquad (153)$$

where R_s is the solid thickness or radius (m). For a slab R_s is the thickness upon total shrinkage after complete drying; for a solid sphere it would be the final radius. Three different curves can be identified from a plot of $\mathbf{F}$ versus X as shown in Figure 4.18:

- The *Regular Regime* (RR) curve is the grandparent curve for a given material, from which all drying curves may be derived.

- The *Penetration Period* (PP) is the parent curve for drying curves for a given initial moisture content at high initial $\mathbf{F}$ values.

- The *Constant Activity* (CA) period is determined by external conditions and initial moisture content.

The flux parameter can be written for the RR period as follows:

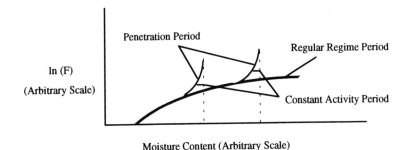

Figure 4.18.   Drying curves using F versus X. (Reprinted from Kerkhof (1994), p. 20 by courtesy of Marcel Dekker, Inc.)

$$F_{RR}(T, X, X_i) \cong F_{RR}(T, X, 0) - F_{RR}(T, X_i, 0)$$
$$= F_{RR}^*(T, X) - F_{RR}^*(T, X_i) \tag{154}$$

where X_i is the water content at the interface, F_{RR}^* is the RR curve for the interfacial $X_i = 0$, and X is the moisture content at a given moment. The dependence on the temperature is described by means of an Arrhenius equation:

$$F_{RR}^*(T, X) = F_{RR}^0(X) \exp\left(-\frac{E_f(X)}{RT}\right) \tag{155}$$

where F_{RR}^0 is a frequency factor, and $E_f(X)$ is the activation energy. The transition between PP and RR is defined as follows:

$$\frac{d(\ln F_{RR})}{dX} = \frac{1}{X_0 - X} \tag{156}$$

and the PP is defined as:

$$F_{PP} = \frac{E_T F_T}{E} \tag{157}$$

$$E = \frac{X_0 - X}{X_0 - X_i} \tag{158}$$

The intersection of the **F** value with either the PP curve or the RR curve then gives the transition of CA to other periods, and thus the complete drying curve is obtained. The R_s value for a layer can be defined as (Kerkhof, 1994):

$$R_s = \frac{L_0 F_{so}}{d_s} = \frac{L_0}{\left(1 + \dfrac{X_0 d_s}{d_w}\right)} \tag{159}$$

where F_{so} is the initial solid concentration, d_s is the specific density of solids, and d_w is the specific density of water. The drying time is given by the following expression:

$$t = d_s R_s^2 \int_X^{X_0} \frac{dX}{F} \tag{160}$$

4.2 THE DRYING PROCESS AND WATER ACTIVITY

The final effect of drying a product is a lower water activity along with a lower moisture content. The steps in which both moisture content and water activity are reduced could be considered as pseudoequilibrium states in which the amount of unbound water is a function of the temperature within the product. If the amount of water within the product is constant, represented by line A in Figure 4.19, an increase in the temperature increases the value of water activity because of the greater amount of energy available for vaporization, as predicted by the Clausius–Clapeyron relationship. The drying process, in most of cases, not only consists of applying heat to the product, which increases the water activity within the product, but also in the removal of water, which results in a reduced moisture content. Thus, considering a drying temperature T_c and an initial water activity a_{wi} the ultimate moisture value in the product will be X_e or the equilibrium moisture content. The X_e is a function of both the temperature of the drying process and the water activity as presented in Figure 4.19.

The drying process of a product with an initial moisture content X_i and a water activity a_{wi}, at a temperature T_a, is described by line C in Figure 4.20. The moisture content will drop while the drying is in progress, but the water activity will remain constant or increase depending on how fast the water vapor is removed from the product. The drying will continue until the equilibrium moisture content is reached at a water activity a_{wh}. At that point the relationship between water activity and the moisture content is defined by the operating temperature, or the sorption isotherm at T_c as stated before. The product cool down leads to a product with a certain moisture content (X_e), and a reduced water activity a_{wc}, based upon the Clausius-Clapeyron relationship, as represented by line A in Figure 4.20.

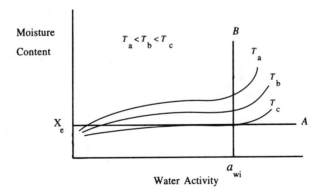

Figure 4.19.   Effect of temperature on water activity.

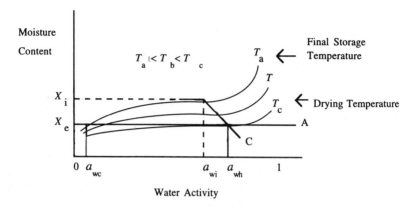

Figure 4. 20.   Drying process and water activity

4.3 CONCLUDING REMARKS

The most relevant theories in drying have been presented in this chapter. The most important aspect to be noted is that the actual driving forces for mass transport are (1) the chemical potential and (2) the temperature gradient within the processed material.

In addition, there are some limitations that must be present when applying any of the models: (1) The heat and mass transfer properties of the biological materials vary with concentration and temperature. The assumption of constant properties is just for convenience to obtain analytical solutions. (2) The shrinkage of materials during drying has to be accounted for because the rate of drying is expressed in terms of surface area. (3) The changes in solute concentration may change the drying curves drastically, and thus change the expected equilibrium moisture content after drying.

The change in water activity and moisture content during drying is an important aspect to consider while drying a foodstuff. The effects of both moisture content and water activity in degradation reactions are also a function of the processing temperature.

4.4 NOMENCLATURE

α	Latent heat coefficient, Whitaker's Theory
β	Constant
β_n	Bessel function roots
δ	Thermal gradient coefficient
δ_p	Thermal gradient due to pressure
ϵ	Porosity or void fraction of solid
ζ	Phase conversion factor of liquid into vapor
ϕ	Isotherm function
Φ	Source of electromagnetic energy
γ	Constant
Γ	$(X - X_s)/(X_o - X_s)$
κ	Permeability
λ	Latent heat (w, water)
μ	Viscosity
η	Dimensionless thermal coefficient
ρ	Density (w, water; l, liquid; s, solid; i, individual, or total) concentration (kg/m^3)
σ	Surface tension
τ	Tortuosity
Ω	Dimensionless mass transfer coefficient
ς	Whitaker function
ψ	Capillary potential or volumetric fraction
A	Area (exposed, s, cross-sectional)
a	Volume fraction of air in pores
a_w	Water activity or relative humidity
b	Saturation level of pore with liquid
Bi	Biot number
c	Specific heat (s, solid; w, water)
C_p	Heat capacity
C_B	Specific body capacity for moisture
c_b	Reduced specific heat of the solid

D_{AB} Molecular diffusivity of air/water mixture

D Diffusion coefficient (m, total; v, vapor; L, liquid; s, surface; eff, effective, o, reference; k, Knudsen diffusivity)

d Diameter

E_a Activation energy

f Void fraction of the body

$f(r)$ Differential curve for pore sizes

F_s Weight of dry solids

$\mathbf{F}$ Flux parameter

G Mass velocity of air

$\mathbf{g}$ Gravitational force

$\mathbf{H}$ Whitaker function

H_c Dimensionless capillary pressure–saturation relationship

$H_{x,T}$ Dimensionless derivatives, Whitaker's Theory

$h_{v,l}$ Enthalpies

h Heat transfer coefficient

h_c Capillary pressure–saturation relationship

h_n 2n + 1

J Flux (L, liquid; V, vapor)

$\mathbf{J}$ Heat or diffusion fluxes

K Mass transfer coefficient (overall, y, moisture; eff, effective)

K_H Unsaturated hydraulic conductivity

$K_{1,2}$ Equilibrium constants

K_T Heat conduction coefficient

K_T Thermal diffusivity (v, vapor; l, liquid; m, total)

k_T Thermal conductivity

k_y External mass transfer coefficient

$\mathbf{K}$ Permeability tensor (s, saturation; <T>, thermal; g, gravitational; β, gas phase; γ, liquid phase)

K_{eff}^t Total thermal conductivity tensor

L Length of plate

L_{em} Modified Lewis number

L_{ik}	Luikov phenomenological coefficient
L_{ik}	Luikov phenomenological coefficient
$L_i(r)$	Luikov phenomenological coefficient
M	Molecular weight (a, water; b, air)
N_a	Flux of water vapor
Nu	Nusselt number
n	Unit normal vector
P	Pressure (total; w, partial of water)
Pr	Prandlt number
q	Heat flux (v, evaporation)
r	Radii
R	Rate of drying (c, constant rate)
Re	Reynolds number
R_g	Gas constant (g, gas; w, water)
T	Temperature (w, wet bulb; ∞, bulk)
t	Time
U	Dimensionless mass transfer parameter
V	Dimensionless heat parameter
v	Mass velocity (average; i, individual)
W	Weight (t, time; s, dry solids)
X	Thermodynamic force (temperature, concentration, etc.)
X	Moisture content in kg of water/kg of dry solids (t, time; o, initial; s, surface; eq, equilibrium; v, as vapor in pores; sm, maximum soprtional liquid content)
X_L	Dimensionless length
x	Spatial dimension
Y	Air moisture content (bulk; w, at surface of solid or wet bulb)

4.5 REFERENCES

Aviles, E. I. 1989. Steady state drying of granular solids. Masters Thesis, Department of Chemical Engineering, University of Puerto Rico, Mayagüez, Puerto Rico.

Berger, D. and Pei, D. C. T. 1973. Drying of hygroscopic capillary porous solids. A theoretical approach. *Int. J. Heat Mass Transfer* 16:293–302.

Bruin, S. and Luyben, K. Ch. A. M. 1980. Drying of food materials. In *Advances in Drying*, Vol. 1, edited by A. S. Mujumdar. Hemisphere Publishing, New York.

Cheftel, J., Cuq, J. L. and Lorient, D. 1985. Amino acids, peptides and proteins. In *Food Chemistry*, Second edition, edited by O. R. Fennema. Marcel Dekker, New York.

Chen, C. S. and Johnson, W. H. 1969. Kinetics of moisture movement in hygroscopic materials. I. Theoretical consideration of drying phenomena. *Trans. ASAE.* 12:109–113.

Chirife, J. 1983. Fundamentals of the drying mechanism during air dehydration of foods. In *Advances in Drying*, Vol. 2, edited by A. S. Mujumdar. Hemisphere Publishing, New York.

Colón, G. and Avilés, E. I. 1993. Transport mechanisms in the drying of granular solids. In *Food Dehydration*, AIChE Symposium Series. 89(297):46–54.

Fortes, M. 1978. A non-equilibrium thermodynamics approach to transport phenomena in capillary-porous media with special reference to drying of grains and foods. Ph.D. Thesis, Purdue University.

Fortes, M. and Okos, M. R. 1980. Drying theories. In *Advances in Drying*, Vol. 1, edited by A. S. Mujumdar. Hemisphere Publishing, New York.

Geankoplis, C. J. 1983. Drying of process materials. In *Transport Processes and Unit Operations*, Second edition, Allyn and Bacon, Boston, MA.

Keey, R. B. 1972. *Drying Principles and Practice.* Pergamon Press, New York.

Keey, R. B. 1978. *Introduction to Industrial Drying Operations.* Pergamon Press, London, UK.

Kerkhof, Piet J. A. M. 1994. The role of theoretical and mathematical modeling in scale-up. *Drying Technol.* 12:1–46.

Luikov, A. V. 1966. Application of irreversible thermodynamic methods to investigation of heat and mass transfer. *Int. J. Heat Mass Transfer* 9:139–152.

Luikov, A. V. 1975. Systems of differential equations of heat and mass transfer in capillary porous bodies (Review). *Int. J. Heat Mass Transfer* 18:1–14.

Okos, M. R., Narsimhan, G., Singh, R. K., and Weitnaver, A. C. 1992. Food dehydration. In *Handbook of Food Engineering*, edited by D. R. Heldman and D. B. Lund. Marcel Dekker, New York.

Philip, J. R. and De Vries, D. A. 1957. Moisture movement in porous materials under temperature gradient. *Trans. Am. Geophys. Union* 38:222–232.

Strongin, V. and Borde, I. 1987. A mathematical model of convective drying incorporating sorption isotherms. In *Advances in Drying*, Vol. 3, edited by A. S. Mujumdar. Hemisphere Publishing, New York.

Toei, R. 1983. Drying mechanism of capillary porous bodies. In *Advances in Drying*, Vol. 2, edited by A. S. Mujumdar. Hemisphere Publishing, New York.

Van Arsdel, N. B. and Copley, M. J. 1963. *Food Dehydration.* AVI Publishing, Westpoint, CT.

Van Brakel, J. 1980. Mass transfer in convective drying. In *Advances in Drying*, Vol. 1, edited by A. S. Mujumdar. Hemisphere Publishing, New York.

Whistler, R. L. and Daniel, J. R. 1985. Carbohydrates. In *Food Chemistry*, Second edition, edited by O. R. Fennema. Marcel Dekker, New York.

Whitaker, S. 1980. Heat and mass transfer in granular porous media. In *Advances in Drying*, Vol. 1, edited by A. S. Mujumdar. Hemisphere Publishing, New York.

Young, J. H. 1969. Simultaneous heat and mass transfer in porous hygroscopic solid. *Trans. ASAE*, 12:720–725.

CHAPTER 5

CABINET AND BED DRYERS

5.0 INTRODUCTION

The main purpose of food dehydration is to extend the shelf life of the final product. The process meets this objective by reducing the moisture content of the product to a level that limits microbial growth and chemical reactions. Hot air is used in most drying operations and air dryers have been in use for several years around the world.

The basic configuration of an atmospheric air dryer is a chamber where the food is placed, and it is equipped with a blower and ducts to allow the circulation of hot air around and across the food. The water is removed from the product surface and carried out from the dryer in a single operation. The air is heated while entering the dryer by means of heat exchangers or direct mixture with combustion exhaust gases. This type of dryer is used widely in the manufacture of cookies, dried fruit and vegetable slices, and pet foods.

Some of the topics discussed in this chapter include types of dryers (i.e., batch and continuous), mass and heat balances, and industrial applications.

5.1 FUNDAMENTALS

In general, the drying phenomenon depends on the heat and mass transfer characteristics of both the drying air and the food product, as discussed in Section 2.3 and Chapter 4. The drying in an atmospheric dryer can be summarized in terms of two phenomena: heating of the product and reduction in moisture content, both as a function of time. Figure 5.1 illustrates the temperature and the moisture profiles as a function of drying time.

Certain types of dryers expose food products to a direct stream of hot air that heat up the product and remove the water vapor. The natures of certain products, however, do not allow a direct exposure to hot air and the heating is accomplished by means of heat exchangers that prevent direct contact between the product and the heat-

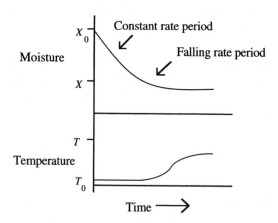

Figure 5.1. Typical temperature and moisture profiles for food dehydration.

Table 5.1. Types of direct dryers as discussed by Porter et al. (1973).

Continuous	Batch
Tray dryer	Air through
Continuous sheet	Tray and compartment
Pneumatic conveying	Fluid bed
Rotary	
Spray	
Through circulation	
Tunnel	
Fluid beds	

ing medium. The first type of dryers are known as *direct dryers* and the second type are known as *indirect dryers* (Cook and DuMont, 1991).

The most commonly found dryer in an atmospheric drying operation is the direct type. Table 5.1 summarizes the 11 types of direct dryers identified by Porter et al. (1973):

5.1.1 Components of a Dryer

The basic configuration of a dryer consists of a feeder, a heater, and a collector. The final arrangement of these components is characteristic for each type of dryer. Figure 5.2 presents a basic scheme for a dryer.

Feeder: The most common feeders for wet solids are screw conveyors, rotating tables, vibratory trays, and rotary air locks (Cook and DuMont, 1991). In some instances special feeders on wide bed dryers are needed to ensure uniform spreading of the material.

Heater: In direct heaters air is heated by mixing it with combustion exhaust gases. In indirect heaters air or product is heated through a heat exchanger. The cost of direct-fired heating is lower than that of indirect, but some products are damaged by the gases (Cook and

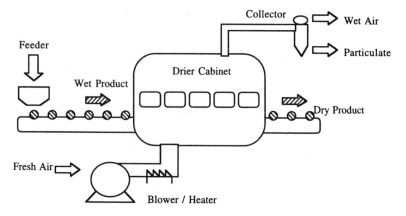

Figure 5.2. Basic configuration of an atmospheric air dryer.

DuMont, 1991). The maximum air temperature possible in a direct heater ranges from 648 to 760°C whereas 425°C is for an indirect heater (Cook and DuMont, 1991).

Collector: The separation of powdered products or product particulates from the air stream can be accomplished using cyclones, bag filters, or wet scrubbers.

5.2 MASS AND HEAT BALANCES

The basics of mass and heat balance for an atmospheric air dryer were discussed in Chapter 2. In this section we discuss specific considerations such as parallel air flow to the drying surface and through flow drying.

5.2.1 Batch Dryers

Air conditions do not remain constant in a compartment or tray dryer while drying occurs. Heat and mass balances are used to estimate the exit-gas conditions (i.e., temperature and humidity), which are summarized in Figure 5.3 for a tray dryer.

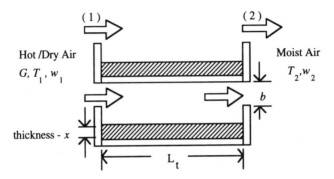

Figure 5.3.   Heat and mass balance variables in a tray dryer.
(Adapted from Geankoplis, 1983. T, temperature; w, air humidity;
G, air flux.)

The heat balance over a length dL_t of a tray for any section with a wide dimension z is expressed as (Geankoplis, 1983):

$$dq = Gc_s(z \cdot b)dT \qquad (1)$$

where b (m) is the distance between trays, z (m) is the width of the tray, G (kg/s m^2) is the dry air flux, q is the heat flux, c_s is the humid heat of the air–water mixture, and T is the temperature. Expressing dq in terms of convective heat transfer:

$$dq = h(z \cdot dL_t)(T - T_w) \qquad (2)$$

where h is the heat transfer coefficient, T_w is the wet bulb temperature, and L_t is the length of the tray. Assuming that h and c_s are constant, Eqs. (1) and (2) can be rearranged and integrated:

$$\frac{hL_t}{Gc_s b} = \frac{T_1 - T_w}{T_2 - T_w} \qquad (3)$$

where T_1 is the inlet temperature and T_2 is the outlet temperature. The log mean temperature difference is defined as:

$$\left(T - T_w\right)_{LM} = \frac{\left(T_1 - T_w\right) - \left(T_2 - T_w\right)}{\ln\left(\dfrac{T_1 - T_w}{T_2 - T_w}\right)} \tag{4}$$

Combining Eqs. (3) and (4):

$$\left(T - T_w\right)_{LM} = \frac{\left(T_1 - T_w\right) - \left(1 - \exp\dfrac{-hL_t}{Gc_s b}\right)}{\left(\dfrac{-hL_t}{Gc_s b}\right)} \tag{5}$$

The constant rate period drying time can be represented as (Geankoplis, 1983):

$$t_c = \frac{x\rho_s L_t \lambda_w \left(X_1 - X_c\right)}{Gc_s b \left(T_1 - T_w\right)\left(1 - \exp\left(\dfrac{-hL_t}{Gc_s b}\right)\right)} \tag{6}$$

where X_1 is the initial moisture content of the product, X_c is the critical moisture content, x is the thickness of the slab, ρ_s is the solid bulk density, and λ_w is the latent heat at T_w. The time expression for the falling rate period can be derived as follows (Geankoplis, 1983):

$$R = \frac{-F_s}{A} \frac{dX}{dt} \tag{7}$$

$$R = \frac{h}{\lambda_w}\left(T - T_w\right) \tag{8}$$

where F_s is the amount of solids. Combining and integrating Eqs. (7) and (8):

$$t = \frac{F_s \lambda_w \left(X_1 - X_2 \right)}{Ah \left(T - T_w \right)} \tag{9}$$

Replacing $(T - T_w)$ by Eq. (5), the falling rate-period time can be expressed as (Geankoplis, 1983):

$$t_f = \frac{x \rho_s L_t \lambda_w X_c \ln \left(\dfrac{X_c}{X} \right)}{Gc_s b \left(T_1 - T_w \right) \left(1 - \exp \left(\dfrac{-hL_t}{Gc_s b} \right) \right)} \tag{10}$$

Equations (6) and (10) take into account the differences between the constant rate and the falling rate periods. The total drying time can be expressed as the sum of both drying times.

5.2.2 Through Circulation Batch Dryer

The drying of packed beds by through circulation is another model in batch drying. The system is shown in Figure 5.4 and can be considered as adiabatic, no heat loss, with a constant air flow G (kg of dry air/h m²) entering at temperature T_1 and moisture content w_1 (kg of H_2O/kg of dry air). Moist air leaves at T_2 and w_2.

The rate of drying and a heat balance on a section dz of the bed can be expressed as:

$$R = G(W_2 - W_1) \tag{11}$$

$$dq = Gc_s A \, dT \tag{12}$$

where A is the cross-sectional area. The heat transfer equation can be expressed as:

$$dq = haA \, (T - T_w) \, dz \tag{13}$$

$$a = 6(1 - \epsilon)/D_p \quad \text{spherical particles} \tag{14}$$

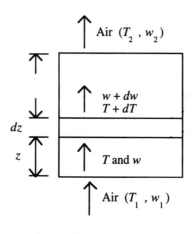

Bed Section

Figure 5.4.   Through circulation drying. (Adapted from Geankoplis, 1983.)

$$a = 4(1 - \epsilon)(1 + 0.5D_c)/(D_c l) \quad \text{cylindrical particles} \quad (15)$$

where l is the length of the particle, D_c is the cylinder diameter, D_p is the diameter of a sphere, and ϵ is the void fraction in the solid. Equating and integrating Eqs. (11) and (12), assuming h and c_s are not temperature dependent:

$$\frac{haz}{Ga_s b} = \ln\left(\frac{T_1 - T_w}{T_2 - T_w}\right) \quad (16)$$

Considering $F_s = A\rho_s/a$, the expression for the drying time for each of the periods (constant and falling) can be expressed as follows (Geankoplis, 1983):

Constant rate period

$$t = \frac{\rho_s \lambda_w (X_1 - X_c)}{hA(T - T_w)} \quad (17)$$

or

$$t = \frac{\rho_s\left(X_1 - X_c\right)}{aK_y M_\beta\left(w_w - w\right)} \tag{18}$$

where K_y is the mass transfer coefficient, M_B is the air molecular weight, and w_w is the air moisture content at T_w.

Falling rate period

$$t = \frac{\rho_s\lambda_w X_c \ln\left(\dfrac{X_c}{X}\right)}{ha\left(T - T_w\right)} \tag{19}$$

or

$$t = \frac{\rho_s X_c \ln\left(\dfrac{X_c}{X}\right)}{aK_y M_B\left(w_w - w\right)} \tag{20}$$

The temperature difference across the bed is approximated using the log mean temperature difference, Equation (4). Then, Eqs. (17) and (19) can be expressed as:

Constant rate period

$$t_c = \frac{x\rho_s\lambda_w\left(X_1 - X_c\right)}{Gc_s\left(T_1 - T_w\right)\left(1 - \exp\left(\dfrac{-hax}{Gc_s}\right)\right)} \tag{21}$$

Falling rate period

$$t_f = \frac{x\rho_s\lambda_w X_c \ln\left(\dfrac{X_c}{X}\right)}{Gc_s\left(T_1 - T_w\right)\left(1 - \exp\left(\dfrac{-hax}{Gc_s}\right)\right)} \tag{22}$$

where x is the thickness of the bed.

The heat transfer coefficient for a through circulation drying can be evaluated as follows (Geankoplis, 1983):

$$h = 0.151 \frac{G_t^{0.59}}{D_p^{0.41}} \qquad \left(D_p G_T / \mu > 350\right) \qquad (23)$$

$$h = 0.214 \frac{G_t^{0.49}}{D_p^{0.51}} \qquad \left(D_p G_T / \mu < 350\right) \qquad (24)$$

The D_p for a cylinder is evaluated as follows:

$$D_p = (D_c l + 0.5 D_c^2)^{0.5} \qquad (25)$$

Example 1

A food product is extruded as cylinders with a diameter of 7 mm and 30 mm length. The initial moisture content is 1.5 kg of H_2O/kg of dry solids. The equilibrium moisture content is 0.01 kg of H_2O/kg of dry solids. The cylinders are packed on a screen to a depth of 60 mm, with a density of 640 kg/m³. The density of the dry solids is 1602 kg/m³. Air flows at 0.811 m/s with an initial moisture content of 0.05 kg of H_2O/kg of dry air and a temperature of 120°C. The critical moisture content is 0.50. Calculate the total time to dry the solids to a moisture content of 0.2 kg of H_2O/kg of dry solids.

Answer

The free moisture content of the product is expressed as follows:

Initial: $X_1 = X_{t1} - X^* = 1.5 - 0.01 = 1.49$ kg of H_2O/kg of dry solids

Critical: $X_c = X_{tc} - X^* = 0.5 - 0.01 = 0.49$ kg of H_2O/kg of dry solids

Final: $X = X_t - X^* = 0.2 - 0.01 = 0.19$ kg of H_2O/kg of dry solids

The incoming air conditions are the following:

$T_1 = 120°C$ $T_w = 49°C$ $w_1 = 0.05$ kg of H_2O/kg of dry air

$w_w = 0.083$ kg of H_2O/kg of dry air $\lambda_w = 2382$ kJ/kg

The humid volume of air and density are evaluated at follows:

Humid volume: $v_H = (2.83E\text{-}3 + 4.56E\text{-}3w)(T + 273)$

At the intake: $v_H = 1.20$ m³/kg of dry air.

The density of the air–water mixture entering the dryer is:

$$\rho = (1 + 0.05)/1.20 = 0.875 \text{ kg of dry air} + H_2O/\text{m}^3$$

The mass velocity of dry air is (base: 1 kg of dry air):

$G = v\rho$

$G = (0.811 \text{ m/s})(0.875 \text{ kg of dry air} + H_2O/\text{m}^3)$
   (kg of air/1.05 kg of air + H_2O)

$G = 0.6758$ kg of dry air/s m² or 2433 kg of dry air/h m²

An average humidity of 0.07 is used to estimate the mass velocity of the air–water mixture, G_t. The air will exit with an humidity of less than 0.083, which is the wet bulb humidity. Thus, an approximation is made by averaging the input humidity and the wet bulb condition.

$$G_t = G(1 + 0.07)$$
$$G_t = 2603 \text{ kg of dry air} + H_2O/\text{h m}^2$$

Also, the humid heat is estimated using the 0.07 kg of H_2O/kg of dry air estimate:

$$c_s = 1.1366 \text{ kJ/kg K}$$

The void fraction, ϵ, is estimated for 1 m³ and the constant a is calculated as follows:

Dry solid mass in 1 m³ is 640 kg, then the volume of solids in the bed is 640/1602 or 0.399 m³. Hence $\epsilon = 1 -$

0.399 or 0.601. D_p is evaluated using Eq. (25) and a using Eq. (15).

$$D_p = 0.015 \text{ m}$$

$$a = 254.6 \text{ m}^2/\text{m}^3$$

The heat transfer coefficient is estimated assuming an average air temperature of 93°C, thus $\mu = 2.15E\text{-}5$ kg/m s or $7.74E\text{-}2$ kg/m h.

$$\text{Re} = D_p G_t/\mu = (0.015)(2603)/(7.74E\text{-}2)$$

$$\text{Re} = 504.45$$

Using Eq. (23) h is 87.49 W/m^2 K. Finally, the drying time can be evaluated as follows:

Constant rate period: Equation (21)

$$t_c = 1800 \text{ s or } 0.51 \text{ h}$$

Falling rate period: Equation (22)

$$t_f = 950 \text{ s or } 0.26 \text{ h}.$$

Total drying time:

$$t = t_c + t_f$$

$$t = 0.77 \text{ h}.$$

Notice from the drying time for each period that the falling rate requires approximately 1.7 times the amount of time needed to remove the same amount of water in the constant rate period.

5.2.3 Continuous Dryers

In Chapter 2 we discussed the heat and mass balance in an ideal counter-current dryer. The equation for the drying time in the constant rate period is expressed as follows (Geankoplis, 1983):

$$t = \left(\frac{G}{F_s}\right)\left(\frac{F_s}{A}\right)\frac{1}{K_y M_B} \ln\left(\frac{w_w - w_c}{w_w - w_1}\right) \tag{26}$$

where A/F_s is the exposed drying surface. Equation (26) can be expressed as:

$$t = \left(\frac{G}{F_s}\right)\left(\frac{F_s}{A}\right)\frac{1}{K_y M_B}\left(\frac{w_1 - w_c}{\nabla w_{LM}}\right) \tag{27}$$

where ∇w_{LM} is the log mean humidity difference:

$$\nabla w_{LM} = \frac{(w_1 - w_w) - (w_c - w_w)}{\ln\left(\dfrac{w_w - w_c}{w_w - w_1}\right)} \tag{28}$$

and w_c is defined as (Geankoplis, 1983):

$$w_c = w_2 + \frac{F_s}{G}(X_c - X_2) \tag{29}$$

The equation for the falling rate period can be expressed as:

$$t = \left(\frac{G}{F_s}\right)\left(\frac{F_s}{A}\right)\frac{X_c}{X_2 + (w_w - w_2)\left(\dfrac{G}{F_s}\right)K_y M_B} \ln\left(\frac{X_c(w_w - w_c)}{X_2(w_w - w_1)}\right) \tag{30}$$

On the case of co-current, as shown in Figure 5.5, the entering hot air contacts the entering wet solid and the mass balance can be expressed as:

$$Gw_1 + F_s X_1 = Gw_2 + F_s X_2 \tag{31}$$

or

$$G(w_2 - w_1) = F_s(X_1 - X_2) \tag{32}$$

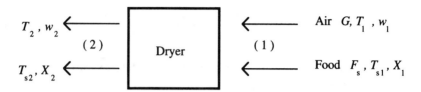

Figure 5.5.   Co-current drying operation.

In terms of critical values:

$$G(w_c - w_1) = F_s (X_1 - X_c) \qquad (33)$$

$$w_c = w_1 + \frac{F_s}{G}\left(X_1 - X_c\right) \qquad (34)$$

5.3 DESCRIPTION OF DRYERS

In this section we describe different types of atmospheric dryers currently used in food processing, as well as their constraints and advantages. The types of dryers to be discussed are the kiln, cabinet or tray, tunnel, and rotary.

5.3.1 Batch Dryers

This type of dryer is used when the operations are small or seasonal, or when different types of materials are dried (Karel, 1975).

5.3.1.1 Kiln Dryer

Grains, fruits, and vegetables can be processed using this type of dryer. It consists of a two-story building with a slotted floor that separates the drying section (in the upper section of the building) from the burners section in the lower floor. Figure 5.6 presents an example of a simple kiln dryer. The product is placed over the slotted floor and the heated air is forced from the lower section to the

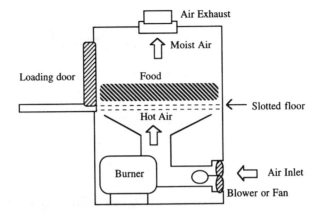

Figure 5.6. Typical configuration of a Kiln dryer. (Adapted from Karel, 1975.)

drying section through the floor. Drying times are quite long due to the extensive amount of product processed per drying cycle. The product rests on the slotted floor while hot air passes through it.

5.3.1.2 Cabinet or Tray Dryer

Trays containing the product are moved into a drying compartment and exposed to the drying air. The air is heated by means of a heater at the entrance and forced through the stack of trays and over the product. The main problem with this type of dryer is obtaining uniform drying at various locations on the drying trays (Heldman and Singh, 1981).

This type of dryer is usually used in small-scale and pilot plant scale operations. The dryer consists of an insulated cabinet, trays, and a heat source for circulating the heated air (Karel, 1975). Air heaters may be direct gas burners, steam coils, exchangers, or electrical heaters. Air velocities of 2 to 5 m/s are used (Brennan et al., 1990). These dryers mainly dry fruits and vegetables at through-

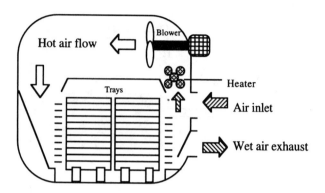

Figure 5.7.   Typical arrangement for a double truck tray dryer. (Adapted from Porter et al., 1973.)

puts of 1000 to 20,000 kg/day. Figures 5.7 and 5.8 illustrate some examples of tray dryers.

5.3.1.3 Rotary Dryer

Rotary dryers are also known as agitated dryers. The housing enclosing the process is stationary, while solids are moved by an internal agitator. Agitated or rotary dry-

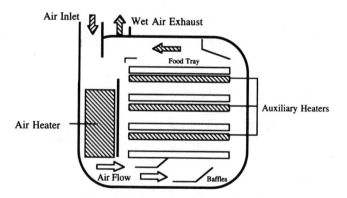

Figure 5.8.   Typical arrangement for a tray dryer with auxiliary heaters. (Adapted from Karel, 1975.).

ers are suitable for processing solids that are free-flowing and granular when discharged as product (Porter et al., 1973). Another type of rotary dryer consists of a double-cone rotating housing as shown in Figure 5.9. The sloping walls of the cones permit more rapid emptying of solids (Porter et al., 1973).

A cylinder mounted over bearings and slightly inclined to the horizontal is also used as rotary dryer. The length of the cylinder may vary from 4 to 10 times its diameter, which may vary from 0.3 m to more than 4 m. Heat transfer may be accomplished by direct or indirect heating. The use of flights in the interior of the cylinder allows the lifting and showering of solids through the drying medium (Porter et al., 1973). Figure 5.10 presents an example of some types of flights currently in use in rotary dryers as well as an example of a dryer.

5.3.2 Continuous Dryers

Continuous tunnel or belt dryers are used in many food dehydration operations. The advantages of a continuous dryer are directly related to the rate of drying and cost

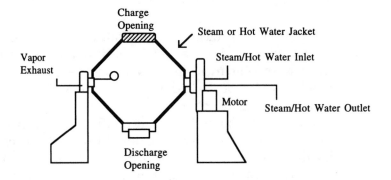

Figure 5.9. Typical arrangement of a double-cone rotary dryer.

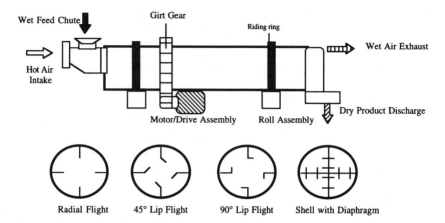

Figure 5.10.   Rotary dryer and flights. (Adapted from Porter et al., 1973.)

effectiveness. Two possible configurations can be used: co-current and counter-current air to product flows. In a co-current operation, the hottest air is in contact with the coolest product when both enter the dryer. This promotes rapid drying of the product at the beginning of the drying operation while the dried product is exposed to a lower temperature.

In a counter-current operation, the hottest air is in contact with the dry product. This situation will promote physical and chemical changes in the dry product such as case hardening because the surface collapses and nonenzymatic browning or scorching occurs.

The selection of the drying configuration is based on the type of product and the desired final properties after drying. The temperature distribution in a dryer varies with the arrangement of air flow and also depends on the drying period, as shown in Figure 5.11.

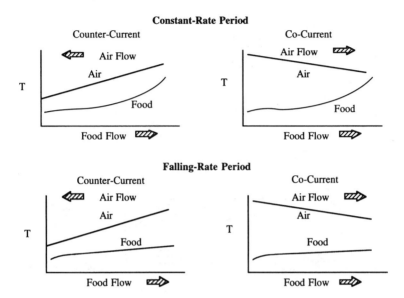

Figure 5.11.   Temperature pattern in a co-current and counter-current drying operation.

5.3.2.1 Tunnel Dryers.

This type of dryer, which may be up to 24 m long with a square or rectangular cross-section about 2 m x 2 m (Brennan et al., 1990), consists of a cabinet equipped with rails to move the tray trucks along the drying chamber. Co-current and counter-current examples are presented in Figure 5.12. Similar to the cabinet dryer, tunnel dryers have nonuniform drying at different locations in the tunnel (Heldman and Singh, 1981). This type of dryer provides a means of drying fruit and vegetables on a semi-continuous basis. An operator loads the trays in trucks and places the truck in the dryer. Then the drying proceeds automatically until the exit, where an operator removes the dry product.

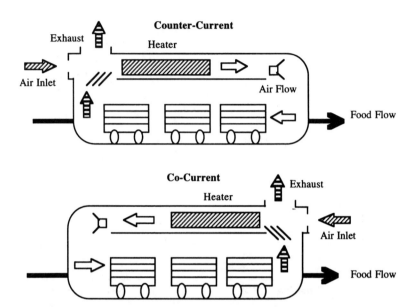

Figure 5.12.   Tunnel dryers. (Adapted from Karel, 1975; and Okos et al., 1992.)

5.3.2.2 Belt or Conveyor Dryers

The principle in a belt dryer is similar to that of a tunnel dryer, except that the product is conveyed through the system on a belt (Brennan et al., 1990; Okos et al., 1992). The most common configuration used in practice is the through-flow. This configuration consists of passing the heated air directly through the belt and layer of product as shown in Figure 5.13. Also, co-current and counter-current configurations can be used as shown in Figure 5.14.

Figures 5.15 through 5.17 present examples of dryers currently in use for processing foodstuffs. Truck, tray, and tunnel dryers (Figure 5.15) are units with a capacity of 1, 2, 4, 8, or 10 trucks. The units are constructed of mild steel, painted with epoxy enamel, or stainless steel. The circulation fan allows air to circulate across trays and

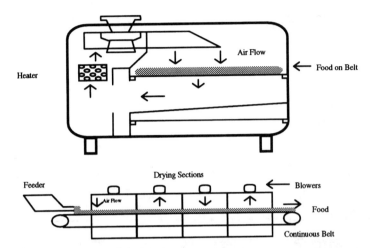

Figure 5.13. Through flow conveyor dryer. (Adapted from Porter et al., 1973; and Okos et al., 1992.)

return to the heater for recirculation. The exhaust provides proper recirculated air characteristics. The heating system will be pneumatic or electric, using gas, steam, electricity, or waste heat as a heating medium. The modular construction design of dryers permits economical additions or extensions of similar units.

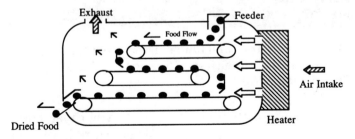

Figure 5.14. Mixed co-current and counter-current belt dryer. (Adapted from Karel, 1975.)

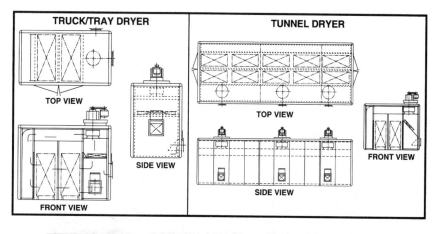

Figure 5.15.   Truck, tray, and tunnel dryers from National Drying Machinery Company®. (Courtesy of National Drying Machinery Company, Philadelphia, PA.)

Figure 5.16 presents a conveyor dryer and cooler from Aeroglide®. The unit's air is controlled by multiple fans, air locks, and internal duct work. The unit is designed with one or two passes with zoned, fully con-

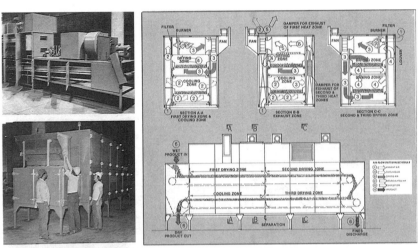

Figure 5.16.   Aeroglide® conveyor dryer and cooler. (Courtesy of Aeroglide Corporation, Raleigh, NC.)

trolled drying. The operation consists of pulling ambient air over a heat source, and alternatively up through and down through the product being dried. Air recirculation usually achieves 70% to 80% efficiency so as to minimize fuel costs and reduce the quantity of exhaust air. Figure 5.17 presents an additional example of a conveyor dryer from National Drying Machinery Company®. The unit consists of a perforated slip-hinge apron conveyor belt. Each zone operates independently, which maximizes throughput while maintaining highest energy efficiency.

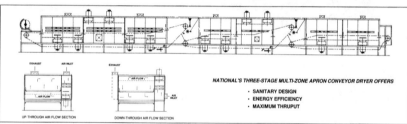

Figure 5.17.   Multiple-stage and single-stage apron conveyor dryers from National Drying Machinery Company®. (Courtesy of National Drying Machinery Company, Philadelphia, PA.)

The heat can be supplied by steam, gas, electric, waste recovery, or a geothermal source.

Table 5.2 includes some examples of products, dryer types, and processing temperatures commonly found in the fruit processing industry. Most applications considered for fruits use hot air at temperatures that range from 60 to 98°C in multiple-stage conveyor dryers while nuts use up to 232°C in a single-stage multiple zone dryer.

Table 5.2.   Typical drying conditions and dryers in fruit processing.

Product	Temperature	Type Dryer
Apples		
Slices, Rings, Dices	60–93°C	Multiple-stage conveyor
Banana Slices	60–82°C	Truck, tunnel
Cherries	60–98°C	Multiple stage conveyor
		Truck tunnel dryers
Coconut flakes	65–93°C	Two stage conveyor
Cranberries	82–93°C	Multiple stage conveyor
Nuts		
Peanuts	65–232°C	Single stage multiple zone dryer
Almonds		
Pistachios		
Pineapple rings and dices	65–93°C	Truck and tunnel dryers

5.3.3 Steam Dryers

Steam drying applications exhibit remarkable energy savings over conventional drying operations (Kumar and Mujumdar, 1990, Potter and Beeby, 1994). The use of superheated steam can be considered in batch ovens, tray dryers, fluidized beds, and pneumatic conveyor dryers.

The drying of agricultural commodities (i.e., cabbage, hay) in superheated steam enables better color retention (Potter and Beeby, 1994). Also, the process eliminates localized overheating of product on hot metallic heat transfer surfaces. This technology allows the production of sterile products with low moisture content.

Some disadvantages of steam drying are the high-temperature operation, which creates problems with temperature-sensitive materials and difficulties in feeding and discharging products arising from the condensation that occurs when cold solids/air is in contact with the steam.

Although the use of steam dryers seems to be an alternative for food processing there is little application

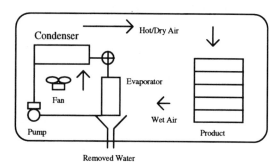

Figure 5.18.   Dehumidification using a heat pump. (Adapted from Heap, 1979.)

in this area. Most of the applications are in drying wood, wood products, paper, and coal as noted by Kumar and Mujumdar (1990).

5.3.4 Heat Pumps

A heat pump extracts heat energy from a source at a low temperature and makes it available as useful heat energy at a higher temperature (Heap, 1979). The use of heat pumps in drying moist materials is an efficient and controllable method. Water vapor can be condensed and the latent heat returned to the drying air via the arrangement shown in Figure 5.18 (Cube and Steimle, 1981; Heap, 1979).

Operating temperatures in dryers using heat pumps range from 50°C to 100°C, considering open-cycle steam recompression units. Heat pump dehumidification can be used in the drying of solid foods and low-temperature concentration of liquids. An example discussed by Heap (1979) is the concentration of orange juice using a heat pump to maintain the juice at 39 to 40°C.

5.4 CONCLUDING REMARKS

Air drying of food products will continue as one of the major drying techniques used at the industrial level. The development of more efficient units, equipped with sophisticated control systems, will follow in the near future. The use of pretreatments to improve organoleptic characteristics (i.e., osmotic dehydration) is taking place within air drying technology and we need to study this as an additional step in developing new products.

5.5 NOMENCLATURE

ϵ	Void fraction of solids
λ_w	Latent heat, kJ/kg of water
ρ_s	Solid bulk density, kg/m^3
μ	Kinematic viscosity, cp
A	Area, m^2
a	Shape factor
b	Spacing between trays, m
c_s	Humid heat of air, kJ/K kg of dry air
D	Diameter (c, cylinder; p, sphere), m
F_s	Amount of solids, kg
G	Air flux, kg/s m^2
h	Heat transfer coefficient, W/m^2
K_y	Mass transfer coefficient, kg/s m^2
LM	Log mean difference
L_t	Total length of a tray, m
l	Length, m
M_B	Molecular weight of air
q	Heat, kJ/s or kW
Re	Reynolds number
T	Temperature (w, wet bulb), °C
t	Time, s or hours

w Humidity of air (1, 2, initial and final; w, wet bulb), kg of water/kg of dry air

X Moisture content (1, 2, initial and final; c: critical, *, equilibrium), kg ofwater/kg of dry solids

x Thickness of product bed or slab, m

z Wide dimension or thickness, m

5.6 REFERENCES

Brennan, J. G., Butters, J. R., Cowell, N. D., and Lilley, A. E. V. 1990. Dehydration. In *Food Engineering Operations*. Third edition. Elsevier Applied Science, New York.

Cook, E. M. and DuMont, H. D. 1991. *Processing Drying Practice*. McGraw-Hill, New York.

Cube, H. L. and Steimle, F. 1981. *Heat Pump Technology*. Butterworth & Co., London, UK.

Geankoplis, C. J. 1983. Drying of process materials. In *Transport Processes and Unit Operations*, Second edition. Allyn and Bacon, Boston, MA.

Heap, R. D. 1979. *Heat Pumps*. Second edition. E. & F. N. Spon, New York.

Heldman, D. R. and Singh, R. P. 1981. *Food Process Engineering*, Second edition. AVI Publishing, Westport, CT.

Karel, M. 1975. Dehydration of foods. In *Principles of Food Science. Part II. Physical Principles of Food Preservation*, edited by M. Karel, O. R. Fennema, and D. B. Lund. Marcel Dekker, New York.

Kumar, P. and Mujumdar, A. S. 1990. Superheated steam drying. *Drying Technol.* 8:195–205.

Okos, M. R., Narsimhan, G., Singh, R. K., Weitnaver, A. C. 1992. Food dehydration. In *Handbook of Food Engineering*, edited by D. R. Heldman and D. B. Lund. Marcel Dekker, New York.

Porter, H. F., McCormick, P. Y., Lucas, R. L. and Wells, D. F. 1973. Gas-solid systems. In *Chemical Engineers's Handbook*, Fifth edition, edited by R. H. Perry and C. H. Chilton. McGraw-Hill, New York.

Potter, O. E. and Beeby, C. 1994. Scale-up of steam-drying. *Drying Technol.* 12:179–215.

=CHAPTER 6

SPRAY DRYING

6.0 INTRODUCTION

Spray drying involves both particle formation and drying,
which makes it a special drying process. The feed is
transformed from the fluid state into droplets and then
into dried particles by spraying it continuously into a hot
drying medium. Similar to fluid bed drying, flash drying,
spray granulation, spray agglomeration, spray reaction,
spray cooling, and spray absorption, spray drying is a suspended particle processing operation (Masters, 1991).
The main differences between spray drying, fluidized bed
drying and flash drying are the feed characteristics (fluid
in spray drying versus solids); residence time (5 to 100 s
for spray drying versus 1 to 300 min for fluidized bed)
and particle size (10 to 500 μm for spray drying versus 10
to 3000 μm for fluidized bed).

The most common spray dryer is the open cycle, co-
current unit illustrated in Figure 6.1 (Dittman and Cook,
1977; Masters, 1991, Shaw, 1994). The open cycle systems have an intake of atmospheric air on a continuous

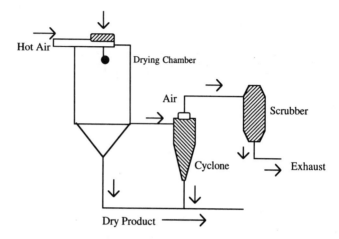

**Figure 6.1.   Open cycle, co-current spray drying layout.
(Adapted from Dittman and Cook, 1977).**

basis. The air is heated, used as drying medium, cleaned
by means of cyclones and scrubbers, and then released
again to the environment. This type of operation presents
a waste of heat contained in the exhaust air. A second
type is the closed loop in which the heating medium (air,
CO_2, etc.) is heated, used in the drying process, cleaned,
dried, and reused again on a continuous basis. The energy
efficiency in this type of dryer is higher than in the open
loop systems. Closed loop systems are more environmen-
tally sound because the output is only the dried product,
whereas open loop systems release hot air and sometimes
microparticulates.

  Spray drying involves the atomization of the feed into
a drying medium which results in moisture evaporation.
The drying proceeds until the desired moisture level in the
product is reached. The drying is controlled by means of
the product and air input conditions (flow and temperature). Finally, the product is recovered from the air.

The advantages of spray drying are the following (Masters, 1991):

- Powder specifications remain constant throughout the dryer when drying conditions are held constant.
- It is a continuous and easy drying operation and adaptable to full automatic control.
- A wide range of dryer designs are available that are applicable to heat-sensitive materials, heat-resistant materials, corrosives, and abrasives.

Installation costs, thermal efficiency, waste heat, and the handling of the exhausting air at saturated or near saturated conditions are the main disadvantages of spray dryers (Masters, 1991).

Spray drying was first applied to the drying of milk shortly after 1900, and was also applied to eggs and coffee in the 1930s. It has been adapted to many other products that are initially liquid and heat sensitive.

6.1 FUNDAMENTALS

The main features of spray drying are the formation of droplets or spray and the contact with air. The atomization step produces a spray for the optimum evaporation condition and subsequently a product within specifications. Atomization results from the breakdown of liquid bulk into small droplets, and the different atomization techniques available vary according to the type of energy used to produce the droplets. The classification of atomizers used in spray drying is summarized in Figure 6.2.

6.1.1 Pressure Nozzles

Pressure nozzles are used to form coarse-particle powders (120 to 300 μm). Variation of pressure gives control over the feed rate and spray characteristics where the latter is

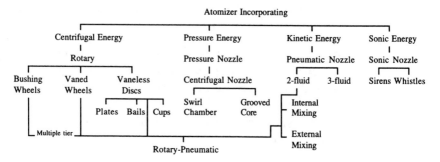

Figure 6.2.   Classification of atomizers. (Adapted from Masters, 1991.)

inversely proportional to pressure. The mean size of the spray is directly proportional to the feed rate and viscosity. Energy transfer is very inefficient during nozzle atomization. Figure 6.3 shows an example of a pressure nozzle.

The conversion of pressure energy within the liquid into kinetic energy of thin moving liquid sheets is the operating principle for this type of nozzle. The liquid sheets break up under the influence of the physical properties of the liquid and by the frictional effects with the air. The power (P_k, kW h/ton of feed) required for a pressure nozzle is proportional to the feed rate and nozzle pressure (Masters, 1991):

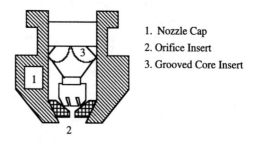

1. Nozzle Cap
2. Orifice Insert
3. Grooved Core Insert

Figure 6.3.   Slotted type pressure nozzle. (Adapted from Masters, 1991.)

$$P_k = 0.27 \frac{\Delta P}{\rho} \qquad (1)$$

where ΔP is the total pressure drop and ρ is the feed density. The conversion of pressure energy into kinetic energy in a centrifugal pressure nozzle results in a rotary motion of the liquid and can be expressed as follows (Marshall, 1954):

$$E_h = 19.2 Q \Delta P \qquad (2)$$

where Q is the flow rate and E_h is the energy or power. The flow of liquid from the orifice of a centrifugal pressure nozzle may be expressed as:

$$2(\pi r_i^2 V_{inlet}) = 2(\pi b r_2 U_r) \qquad (3)$$

$$V_{inlet} = \frac{Q_l}{2\pi r_i^2 \rho} \qquad (4)$$

or

$$\frac{U_r}{V_{inlet}} = \frac{r_1^2}{r_2 b} \qquad (5)$$

where b is the liquid film thickness at the orifice, r_1 is the inlet feed channel radius, r_2 is the orifice radius, V_{inlet} is the inlet velocity of liquid, U_r is the vertical component of velocity of the spray, and Q_l is the mass liquid flow rate. The velocity of the liquid leaving the nozzle is expressed as:

$$V_{res} = (U_h^2 + U_v^2)^{0.5} \qquad (6)$$

or expressed in terms of the pressure drop across the nozzle:

$$V_{res} = C_v (2gh)^n = C_v \left(2g \frac{\Delta P}{\rho} \right)^n \qquad (7)$$

where U_h and U_v are the horizontal and vertical velocity components, $n = 0.5$ for turbulent flow, C_v is a velocity coefficient, **g** is the gravitational force, n is a constant, and h is the pressure head. The performance of a pressure nozzle is affected by the pressure, liquid density, and viscosity. The following approximation was proposed by Masters (1991) to correlate changes in flow rate across the nozzle to changes in either pressure or density:

$$\frac{Q_2}{Q_1} = \left(\frac{P_2}{P_1}\right)^{0.5} = \left(\frac{\rho_1}{\rho_2}\right)^{0.5} \tag{8}$$

The effect of viscosity on flow rate cannot be clearly defined for the pressure nozzle. The exact effect must be determined by experiments for a given operational condi-

Table 6.1.   **Effect of some process variables on droplet size.**

Variable	Effect
Nozzle capacity	
feed rate below design	Atomization incomplete
feed rate at minimum specifications	Decrease droplet size
feed rate within specifications	Increase droplet size
Spray angle	
large angles	Small droplets
Pressure	
increase (medium range)	Decrease size
Viscosity	
increase	Coarse atomization
	Droplet size varies with $\eta^{0.17}$ to $\eta^{0.2}$
very high	Not suitables
Surface tension	
high	Difficult to atomize
Orifice size	Drop size $= kD^2$
	D is the orifice diameter
	k is a constant

Adapted from Masters, 1991.

tion. The effects of process variables such as nozzle capacity, spray angle, pressure, viscosity, surface tension, and orifice diameter on droplet size are summarized on Table 6.1.

Industrial dryers, in which multi-nozzle assemblies are installed to handle the high feed rate, are designed to provide equal operational conditions at each nozzle for uniformity of feed atomization. The nozzle's configuration should meet the following requirements: easy access and removal of nozzles, uniformity of distribution, possibility of isolation, and visibility of each nozzle. Some possible configurations are shown in Figure 6.4.

6.1.2 Rotary Atomizers

Rotary atomizers differ from pressure nozzles in that the liquid attains its velocity without high pressure. Also, the feed rate can be controlled with discs, whereas in the case of nozzles both pressure drop and orifice diameter have to be changed simultaneously. Figure 6.5 shows that physical properties of the feed product control the mechanism of atomization, in this case for a vaneless disc. The formation and release of droplets from the edge of the

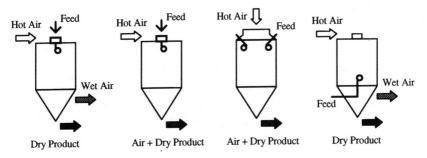

Figure 6.4.   Configuration of pressure nozzle in industrial dryers. (Adapted from Masters, 1991.)

disc, considering low feed rate and disc speed, are shown in Figure 6.5a. The spray consists of a parent droplet and two satellites. An increase in disc speed and feed rate causes a change from the direct-droplet mechanism to one of ligament breakup as shown in Figure 6.5b. A liquid sheet (Figure 6.5c) forms when the liquid ligaments join each other, and then extends from beyond the edge of the disc. This phenomenon is also referred to as the velocity spraying mechanism (Masters, 1991).

On vane wheels (Figure 6.6a) the liquid disintegration occurs at the wheel edge due to the frictional effect between air and the fluid surface. The liquid emerges as a thin film from the vane. The optimum droplet size for a given feed rate depends on the following conditions (Masters, 1991): vibrationless rotation; large centrifugal force; smooth and complete wetting of vane surface; and uniform feed rate and distribution.

Acceleration along the vane ceases when the liquid reaches the edge of the wheel, thus the radial velocity of the liquid can be expressed as follows (Masters, 1991):

$$U_r = 0.0024 \left(\frac{\rho \pi^2 N^2 dQ^2}{\mu h^2 n^2} \right)^{0.33} \tag{9}$$

$$U_t = \pi dN \tag{10}$$

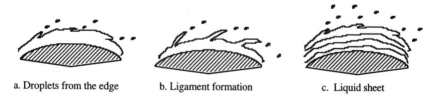

a. Droplets from the edge   b. Ligament formation   c. Liquid sheet

Figure 6.5.   Atomization from a vaneless disc. (Adapted from Masters, 1991.)

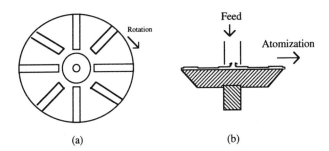

Figure 6.6.   Rotary atomizers: (a) vaned disc (adapted from Masters, 1991); and (b) sharp-edge flat disc (adapted from Heldman and Singh, 1981).

$$U_{res} = [U_r^2 + U_t^2]^{0.5} \tag{11}$$

$$\alpha = \tan^{-1}(U_r/U_t) \tag{12}$$

where U_r is the radial component of velocity, U_t is the tangential component of velocity, U_{res} is the resultant release velocity, α is the angle of liquid release, D is the wheel diameter, N is the rotation speed of the atomizer, n is the number of vanes, h is the height of the vanes, and μ is the viscosity. The effects of process variables such as wheel speed, feed rate, liquid viscosity, surface tension, and liquid density on droplet size from a rotary atomizer are summarized in Table 6.2.

The mean size of droplets can be estimated by the following mathematical models (Masters, 1991):

$$D_{vs} = 140,000[(\mu Q_l/d)^{0.2}(\sigma/nh)^{0.1}]/(N^{0.6}\rho^{0.5}) \tag{13}$$

$$D_{vs} = 114,000r(Q_p/\rho Nr^2)^{0.6}(\mu/Q_p)^{0.2}(\sigma\rho nh/Q_p^2)^{0.1} \tag{14}$$

$$D_{vs} = 14,000\, Q_l^{0.24}/[(nh)^{0.12}(Nd)^{0.6}] \tag{15}$$

where D_{vs} is the Sauter mean diameter (m), d is the atomizer wheel diameter (m), h is the vane height (m), N is the velocity (rpm), Q_l is the mass liquid feed rate (kg/h), Q_p is

Table 6.2.   Effect of process variables on droplet size from a rotary atomizer.

Variable	Effect	
Wheel speed Constant feed rate	$\dfrac{D_1}{D_2} = \left[\dfrac{N_2}{N_1}\right]^p$	D = wheel diameter N = wheel speed, rpm p = 0.55–0.80
Feed rate Constant wheel speed	$\dfrac{D_1}{D_2} = \left[\dfrac{Q_1}{Q_2}\right]^q$	Q = feed rate q = 0.1–0.12
Liquid viscosity	$\dfrac{D_1}{D_2} = \left[\dfrac{\mu_1}{\mu_2}\right]^r$	μ = viscosity r = 0.2
Surface tension	$\dfrac{D_1}{D_2} = \left[\dfrac{\sigma_1}{\sigma_2}\right]^s$	σ = Surface tension s = 0.1–0.5
Liquid density	$\dfrac{D_1}{D_2} = \left[\dfrac{\rho_2}{\rho_1}\right]^t$	ρ = Density t = 0.5

From Masters, 1991.

the vane liquid loading rate (kg/h m), ρ is the density (kg/m³), μ is the viscosity (cP), and σ is the surface tension (dynes/cm). Equation (15) offers the best procedure for predicting droplet and particle size in industrial dryers (Masters, 1991).

The size distribution of sprays from a rotary atomizer can be represented as follows:

$$D_{\text{mean}} = \frac{KQ^a}{N^b d^{0.6}(nh)^d} \tag{16}$$

where the values K, a, b, and d are a function of the wheel speed and vane loading rate. Table 6.3 summarizes the reported values.

Table 6.3. Power and *K* values for Eq. (16).

Wheel Speed (m/s)	Vane Loading Rate (kg/h m)	*a*	*b*	*d*	*K* * 10^{-4}
Normal	Low				
		0.24	0.82	0.24	1.4
85–115 Normal–high	250 Normal				
		0.2	0.8	0.2	1.6
85–180 Very high	250–1500 Normal–high				
		0.12	0.77	0.12	1.25
180–300 Normal–high	1000–3000 Very high				
		0.12	0.8	0.12	1.2
85–140	3000–60,000				

From Masters, 1991.

Rotary atomizers are usually installed at the center of the roof in spray dryrs to provide sufficient contact time to allow evaporation of the liquid (Shaw, 1994).

The relationship between the wet droplet size and the dry particle size can be expressed as follows:

$$D_{wet} = \beta D_{dry} \tag{17}$$

where D_{wet} is the droplet size on atomization, D_{dry} is the dry particle size, and β is a shape-change factor. The factor β is a function of the type of product and temperature and is useful when scaling up a dryer (Masters, 1991).

6.1.3 Pneumatic Atomizers—Two Fluids

The atomization of liquid bulk using high velocity gas is known as pneumatic atomization. The mechanism involves using high-velocity gas to create high frictional forces and this causes the liquid to break down into droplets. Droplet formation occurs in two steps: first, the tearing of the liquid into filaments and large droplets; second, the liquid filaments and big droplets are broken down into smaller droplets. The process is affected by the liquid properties: surface tension, density, and viscosity; and the gas flow properties: velocity and density.

Pneumatic atomization uses as primary gaseous media air and steam. Inert gases are chosen for specialized closed cycle spray systems. The process conditions for optimum frictional conditions require high relative velocity between liquid and air. These conditions are obtained by expanding the gaseous phase to sonic or supersonic velocities before it is in contact with the liquid or by directing the gaseous flow onto a thin liquid sheet within the nozzle (Masters, 1991). Pneumatic nozzles include internal mixing, external mixing, and combined internal/external mixing. The power requirement for an isentropic expansion is expressed as (Masters, 1991):

$$P = 0.402 M_A T \left\{ 0.5 M_a^2 + 2.5 \left[1 - \left(\frac{P_1}{P_2} \right)^{0.286} \right] \right\} \tag{18}$$

where M_A is the mass flow of air (kg/s), T is the air temperature (K), M_a is the Mach number, and P_1 and P_2 are the initial and final pressures. The mean spray size obtained from a pneumatic atomizer can be expressed as follows:

$$D = \frac{A}{\left(V^2 \rho_a \right)^\alpha} + B \left(\frac{M_{air}}{M_{liq}} \right)^{-\beta} \tag{19}$$

where V is the relative velocity between air and liquid, α and β are functions of the nozzles, A and B are constants, M_{air} is the air mass flow, and M_{liq} is the liquid mass flow.

Table 6.4. Effect of process variables on droplets from pneumatic atomizers.

Variable	Effect
Mass ratio air/liquid	
Increase ratio	Decrease droplet size
$M_{air}/M_{liq} < 0.1$	Deteriorative atomization
$M_{air}/M_{liq} \geq 10$	Upper limit for effective ratio
	Increase to create smaller size particles
Relative velocity	
Increase air velocity	Decrease droplet size
Viscosity	
Increase fluid viscosity	Increase droplet size
Increase air viscosity	Decrease droplet size

From Masters, 1991.

Table 6.4 summarizes the effects of process variables on droplets from a pneumatic atomizer.

6.2 INTERACTION BETWEEN DROPLETS AND DRYING AIR

The distance traveled by a droplet until it is fully affected by the air flow depends upon the droplet size, shape, and density. Coarse sprays are more independent of the air flow while fine sprays are considered to move under the influence of the air flow (Masters, 1991).

The movement of the spray is classified according to the dryer layout as co-current, counter-current, or mixed flow as shown in Figure 6.7.

The spray movement can be explained with reference to a single droplet. The forces acting on a droplet can be expressed as follows:

$$\frac{\pi}{6} D^3 \frac{dv}{dt} = \frac{\pi}{6} D^3 \left(\rho_w - \rho_a \right) g - 0.5 C_d \rho_a V_r^2 A \qquad (20)$$

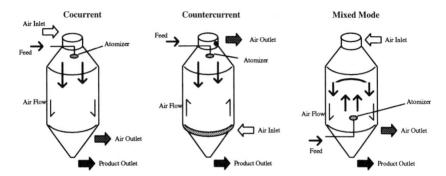

**Figure 6.7.   Classification of dryers according to spray movement.
(Adapted from Heldman and Singh, 1981; Masters, 1991; Shaw, 1994.)**

where D is the droplet diameter, C_d is the drag coefficient,
V_r is the droplet velocity relative to air, A is the droplet
area $(\pi D^2/4)$, ρ_w is the density of the droplet and ρ_a is the
density of the air. Masters (1991) discusses, in more detail,
the movement of spray under different flow conditions.

The temperature profile within a spray dryer is
another important aspect that is a function of the flow
pattern as shown in Figure 6.8.

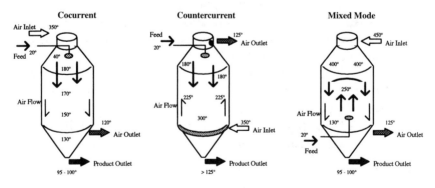

**Figure 6.8.   Temperature profiles in spray dryers. (Adapted from
Masters, 1991.)**

6.3 HEAT AND MASS BALANCES

In the spray drying of food products the liquid to be removed is almost always water, although the removal of organic solvents in closed cycle operations is becoming widespread. The latter is outside the scope of this book. Based on Figure 6.9 both heat and mass balances are drawn up as follows:

A mass balance around the system based upon the moisture with no accumulation leads to:

$$F_s w_{s1} + G_a H_{a1} = F_s w_{s2} + G_a H_{a2} \qquad (21)$$

or

$$F_s(w_{s1} - w_{s2}) = G_a(H_{a2} - H_{a1}) \qquad (22)$$

where F_s is the dry solid rate, w_{s1} is the moisture content of solid entering the dryer, w_{s2} is the moisture content of solid leaving the dryer, G_a is the dry air rate, H_{a1} is the air

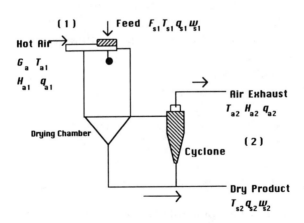

Figure 6.9.   Dryer data for heat and mass balances (F_s, dry solid rate; T_s, temperature of solids; q_s, enthalpy of solids; w_s, moisture content of solids; G_a, dry air rate; T_a, air temperature; H_a, air humidity; q_a, air enthalpy.)

moisture content entering the dryer, and H_{a2} is the moisture content of air leaving the dryer.

The heat or enthalpy balance leads to:

$$F_s q_{s1} + G_a q_{a1} = F_s q_{s2} + G_a q_{a2} + q_L \qquad (23)$$

where q_{s1} and q_{s2} are the enthalpies of solid entering and leaving the dryer, q_{a1} and q_{a2} are the enthalpies of air entering and leaving the dryer, and q_L is the heat loss.

The performance of spray drying is measured in terms of thermal efficiency, which is related to the heat input required to produce a unit weight of dried product of desired specifications. The overall thermal efficiency ($\eta_{overall}$) is defined as the fraction of total heat supplied to the dryer used in the evaporation process:

$$\eta_{overall} = 100 \frac{\left(T_1 - T_2\right)}{\left(T_1 - T_0\right)} \qquad (24)$$

where T_1 is the entering temperature of hot air, T_2 is the exiting temperature of air and T_0 is the atmospheric air temperature. The evaporative efficiency (η_{evap}) is defined as the ratio of the actual evaporative capacity to the capacity obtained in the ideal case of exhausting air at saturation temperature:

$$\eta_{evap} = 100 \frac{\left(T_1 - T_2\right)}{\left(T_1 - T_{sat}\right)} \qquad (25)$$

6.4 DRYING OF DROPLETS

The most important step in a spray drying operation is the droplet formation, which has already been covered in Section 6.1. The exchange of heat and mass starts just after the droplet is released from the atomizer and continues while this droplet travels in the hot air. The droplet, which consists of water and solids in the same proportion as that in the liquid feed, begins to lose water in the form of vapor, and thus begins the formation of a particle. At the end of the drying path within the spray dryer the particle is completely formed, mainly with solids.

6.4.1 Pure Liquid Droplets

Based on the assumption that the moisture removal from a droplet occurs at a constant rate, evaporation can be described by treating the droplet as pure liquid (Heldman and Singh, 1981; King et al., 1984). The heat for evaporation is transferred by conduction and convection to the droplet while it travels into the air stream, and water vapor is transferred by diffusion and convection from the droplet. The rate of heat and mass transfer is a function of air temperature, humidity, droplet diameter, droplet temperature, relative velocity, and type of solids in the liquid.

The mass and heat transfers ocur under turbulent conditions around the liquid droplet (Dlouchy and Gauvin, 1960; Crowe, 1980; Heldman and Singh, 1981; Masters, 1991) and can be expressed as follows:

$$\text{Nu} = \frac{hl}{k} = 2 + K_1 \text{Re}^{0.5} \text{Pr}^{0.33} \tag{26}$$

$$\text{Sh} = \frac{k_y l}{D_{\text{eff}}} = 2 + K_2 \text{Re}^{0.5} \text{Sc}^{0.33} \tag{27}$$

where Nu is the Nusselt number, Sh is the Sherwood number, Re is the Reynolds number, Pr is the Prandtl number v/α or $C_p v/k$, C_p is the heat capacity, α is the thermal diffusivity, v is the kinematic viscosity, Sc is the Schmidt number $v/1$, h is the heat transfer coefficient, l is a characteristic length, k_y is the mass transfer coefficient, D_{eff} is the diffusivity, and $K_{1,2}$ are constants reported as 0.6 (Masters, 1991).

The relation between force and velocity, a critical factor during drying, is given as follows:

$$F = 0.5 C_d U_r^2 \rho_g A \qquad (28)$$

$$C_d = \frac{24}{Re}\left(1 + 0.15 Re^{0.687}\right) \qquad \left(Re < 1000\right) \qquad (29)$$

$$C_d = \frac{24\mu}{U_r \rho_g d} \qquad \left(Re < 1\right) \qquad (30)$$

where C_d is the drag coefficient, μ is the viscosity, A is the surface area, U_r is the relative velocity of the particles with respect to air, d is the droplet diameter, and ρ_g is the air density. The heat and mass balance on the droplet leads to the following expressions (Heldman and Singh, 1981):

Mass balance

$$\frac{dw}{dt} = K_m A\left(p_w - p_a\right) = \frac{2\pi D_{eff}\rho_l d\left(p_w - p_a\right)}{\rho_g} \qquad (31)$$

or by introducing the ideal gas law (Charm, 1978):

$$\frac{dw}{dt} = \frac{2\pi D_{eff} M_w d\left(p_w - p_a\right)}{R_g T_a} \qquad (32)$$

Heat balance

$$\frac{dw}{dt} = \frac{hA}{\lambda}\left(T_a - T_w\right) = \frac{2\pi K_g d}{\lambda}\left(T_a - T_w\right) \qquad (33)$$

$$h = 2K_f/d \qquad \text{for } Re < 20 \qquad (34)$$

where w is the moisture, t is the time, K_m is the mass transfer coefficient, p_w is the water partial pressure, p_a is the air partial pressure, D_{eff} is the diffusivity, ρ_l is the liquid density, ρ_g is the gas density, d is the droplet diameter, M_w is the molecular weight of water, R_g is the universal gas constant, T_a is the temperature of air, h is the heat transfer coefficient, λ is the latent heat of vaporization, K_g is the thermal conductivity of air, T_w is the wet bulb temperature, and K_f is the film thermal conductivity. Integrating Eq. (27) considering the boundary conditions $t = 0$, $d = d_o$ and $t = \infty$, $d = 0$ leads to the following relationship:

$$t_{cr} = \frac{\rho_1 \lambda}{2\Delta T} \int \frac{dd}{h} = \frac{\left(\rho_1 \lambda d_o^2\right)}{8K_g\left(T_a - T_w\right)} \tag{35}$$

where t_{cr} is the drying time at constant rate, d_o is the initial droplet diameter, the upper limit on the integral is 0 because the droplet disappears (pure liquid, no solids). Equation (35) can be expressed in terms of a finite final diameter as follows (Charm, 1978):

$$t_{cr} = \frac{\lambda\left(\rho_1 \lambda d_1^2 - \rho_2 d_2^2\right)}{8K_g\left(T_a - T_w\right)} \tag{36}$$

where $\rho_1 d_1$ is the initial condition and $\rho_2 d_2$ is the final condition during drying. Equations (35) and (36) apply only to low Reynolds number of stationary droplets (Heldman and Singh, 1981).

EXAMPLE 1 (ADAPTED FROM HELDMAN AND SINGH, 1981)

Determine the drying time considering both a *pure liquid droplet* (Eq. 35) and a *finite final diameter droplet* (Eq. 30) while drying a food with a liquid density of 1000 kg/m³. The thermal conductivity (K_g) is 0.04 W/m K and air temperature is 120°C with 0.02 kg of water/kg of dry air. The initial droplet diameter is 20 μm and the final droplet diameter is 10 μm. Assume stationary conditions for the droplets.

Solution

The wet bulb temperature is obtained from the psychrometric chart, $T_w = 42°C$, and the latent heat is 2202.59 kJ/kg. Using Eq. (29) for a *pure liquid droplet*:

$$t_{cr} = (\rho_l \lambda d_0^2)/[8K_g(T_a - T_w)]$$

$$t_{cr} = \frac{\left(1000 \text{ kg/m}^3\right)\left(2202.59 \text{ kJ/kg1}\right)\left(20E - 6\right)^2\left(1000 \text{ J/kJ}\right)}{8\left(0.04 \text{ W/mK}\right)\left(120 - 42\right)}$$

$$t_{cr} = 0.035 \text{ s}$$

Eq. (30) is used for the *finite final diameter* condition:

$$t_{cr} = \frac{\lambda\left(\rho_1 d_1^2 - \rho_2 d_2^2\right)}{8K_g\left(T_a - T_w\right)}$$

The change in density can be evaluated from the change in volume and the water loss during drying. The initial and final volumes are evaluated for a sphere as

$4\pi r^2/3$: $V_i = 4.19E\text{-}15\text{m}^3$ and $V_f = 5.24E\text{-}16 \text{ m}^3$.

The water loss, evaluated from the change in volume and considering a water density of 943.5 kg/m^3, is 3.458E-12 kg of water. Then, the final density is evaluated as follows:

$$\rho_2 = \frac{\left(\text{Droplet mass}\right)_i - \text{Water loss}}{\left(\text{Droplet volume}\right)_i - \text{Volume change}}$$

$$\rho_2 = \frac{\left[\left(4.19E - 15 * 1000\right) - 3.46E \text{-} 12\right]\text{kg}}{\left[4.19E - 15 - 3.67E \text{-} 15\right]\text{m}^3}$$

$$\rho_2 = 1394.28 \text{ kg/m}^3.$$

Replacing all the values in Equation (30):

$$t_{cr} = \frac{(2202.59 \text{ kJ/kg}) \left[1000 \text{ kg/m}^3 (20E - 6\text{m})^2 - 1394.28 \text{ kg/m}^3 (10E - 6\text{m})^2 (1000 \text{ J/kJ}) \right]}{8(0.04 \text{ W/mK})(120 - 42)}$$

$t_{cr} = 0.023\text{s}.$

The time for completing the constant rate period is a function of moisture content when the particle reaches a constant diameter. The critical moisture content of the particle is achieved at the end of the constant rate period. A heat balance during the falling rate period is expressed as follows:

$$\rho_p V_p \lambda \frac{dw}{dt} = hA(T_a - T_w) + \rho_p V_p c_p \frac{dT}{dt} \tag{37}$$

or

$$\frac{dw}{dt} = \frac{hA(T_a - T_w)}{\rho_p V_p \lambda} + \frac{c_p}{\lambda} \frac{dT}{dt} \tag{38}$$

where ρ_p is the droplet density, V_p is the droplet volume and c_p is the specific heat of the droplet. Equation (38) can be integrated by eliminating the last term to obtain an expression for the drying time during the falling rate (Heldman and Singh, 1981):

$$t_{fr} = \frac{(\rho_p d_c \lambda)(w_c - w_c)}{(6h\Delta T_{ave})} \tag{39}$$

where d_c is the droplet diameter at w_c, w_c is the critical moisture content, w_e is the equilibrium moisture content, and ΔT_{ave} is the average temperature difference between the particle and air during the drying period.

EXAMPLE 2 (ADAPTED FROM HELDMAN AND SINGH, 1981)

Estimate the drying time in the falling rate period, with the following conditions:

Food product: Initial total solids of 10%
Density—liquid: 1000 kg/m³
Initial droplet volume: 4.19E-15 m³
Water loss during constant rate period:
3.458E-12 kg
Final moisture content: 0.05 kg of
water/kg of solids
Density-particle: 1394.3 kg/m³

Drying air: $T_a = 120°C$
$T_w = 42°C$
$K_g = 0.04$ W/mK
$\lambda_w = 2202.59$ kJ/kg

Solution

A unit volume of the food contains 900 kg of water/m³ and 100 kg of solids/m³. The initial mass of water and solids in a droplet is calculated from the initial volume of the droplet: $m_{water} = 3.771E$-12 kg and 4.19E-13 kg. Since the droplet weight loss is 3.458E-12 kg during the constant rate period, the droplet has 3.13E-13 kg water at the beginning of the falling rate period. The moisture content, w_c, is calculated as follow:

$$w_c = 3.13E\text{-}13 \text{ kg of water}/4.19E\text{-}13 \text{ kg of solids}$$

$$w_c = 0.747 \text{ kg of water/kg of solids}$$

The average temperature and the heat transfer coefficient are evaluated as follows:

$$\Delta T_{ave} = (T_a - T_w)/2$$
$$= (120 - 42)/2$$
$$= 39°C$$
$$h = 2\,K_g/d$$
$$= 2\,(0.04)/10E\text{-}6$$
$$= 8000 \text{ W/m}^2 \text{ K}$$

Finally, the drying time can be evaluated using Eq. (39):

$$t_{fr} = \frac{(\rho_p d_c \lambda)(w_c - w_e)}{(6h\Delta T_{ave})}$$

$$t_{fr} = \frac{(1394.3 \text{ kg/m}^3)(10E - 6m)(2202.59 \text{ kJ/kg})(0.747 - 0.05)(1000 \text{ J/kJ})}{6(8000 \text{ W/m}^2\text{K})(39)}$$

$$t_{fr} = 0.1114\text{s}.$$

The total drying time is obtained by adding Eqs. (35) and (39) for a droplet in static air (Re ~ 0 , Heldman and Singh, 1981):

$$t = \frac{\rho_l \lambda d_0^2}{8K_g(T_a - T_w)} + \frac{(\rho_p d_c \lambda)(w_c - w_e)}{(6h\Delta T_{ave})}$$

$$= 0.023 + 0.114$$

$$= 0.137 \text{ s} \tag{40}$$

6.4.2 Droplets with Dissolved Solids

The rate of evaporation in droplets containing dissolved solids is lower than that in pure liquid droplets. A decrease in the vapor pressure on the liquid is caused by the presence of the solids (Masters, 1991). An increase in the droplet temperature above the wet bulb temperature is caused by the decrease in vapor pressure. The effect of dissolved solids on vapor pressure is illustrated in Figure 6.10.

The formation of solid structures changes the subsequent evaporation process and Equation (38) is expressed as:

$$\frac{dw'}{dt} = \frac{dw}{dt} * \text{weight of dry solids} \tag{41}$$

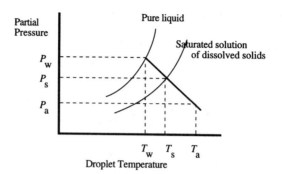

Figure 6.10.   Effect of dissolved solids on droplet temperature (T_w, wet bulb temperature; T_s, droplet surface temperature; T_a, air temperature). (Adapted from Masters, 1991).

or

$$\frac{dw'}{dt} = \frac{-12 K_d \Delta T}{\left(\rho_s \lambda d_c^2\right)} \tag{42}$$

where w' is the amount of water contained by the solids, and K_d is the thermal conductivity of the drying medium. An increase in the solid phase causes a reduction on the moisture movement (an increase in the resistance to mass transfer). The particle begins to heat up because the rate of heat transfer becomes predominant and subsurface evaporation occurs when the heat transfer is high enough to cause vaporization within the droplet. Vapor will be released either by porous formation or rupture of the crust. Figure 6.11 shows the effect of crust formation on the evaporation rate during spray drying.

6.4.3 Droplets with Insoluble Solids

Slurries and pastes are examples of products with insoluble solids which are said to have a negligible effect on the

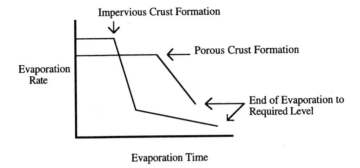

**Figure 6.11.   Effect of crust properties on drying rate.
(Adapted from Masters, 1991.)**

water vapor pressure. The temperature of the droplet can be expressed as equal to the wet bulb temperature of pure liquid droplets during the constant rate period. Equations (36), (39), and (40) can be used to represent the drying times. The ΔT_{ave} is conveniently taken as the logarithmic mean temperature difference (LMTD) between the inlet air temperature, the product feed temperature, the exhaust air temperature, and the droplet surface temperature.

6.5 MICROSTRUCTURE OF SPRAY-DRIED PRODUCTS

The shape characteristics of spray-dried products depend on whether the drying air temperature is above or below the boiling point of the droplets, as shown in Figure 6.12. Particles can be porous rigid, rigid with fractures, pliable, nonporous plastic, and spongelike or crystal "fur" growth, among others.

The bulk density of a spray-dried product can be increased with an increase in the feed rate, temperature of difficult atomizable products, temperature of the powder, solid content, the outlet air temperature, the atomization through a rotary atomizer, or the use of counter-current

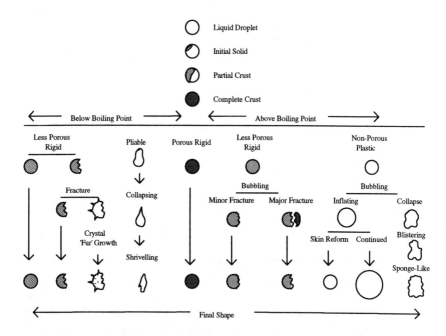

Figure 6.12.   Characteristics of droplets undergoing drying.
(Adapted from Charlesworth and Marshall, 1960. Reproduced with
permission of the American Institute of Chemical Engineers. © 1960
AIChE. All rights reserved).

configuration. Meanwhile, the bulk density will decrease
with an increase in temperature of easy atomizable feed,
increase in inlet air temperature, coarse homogeneous
particle size atomization, feed aeration, or the use of co-
current configuration.

6.6 RECONSTITUTION OF SPRAY-DRIED PRODUCTS

Instant properties of spray-dried products involve the
ability of the powder to dissolve in water. Each particle
has to be wetted, sunk in the liquid, and dissolved. Prop-

erties such as wettability, sinkability, dispersibility, and solubility are of importance for the reconstitution process. Pisecky (1986) and Masters (1991) define these properties as follows:

Wettability	This is the ability of a powder particle to adsorb water on its surface. It is a functional property of powders and is affected by the agglomeration process, the amount of available wettable, surface or the reasonable absence of nonagglomerated particles.
Sinkability	This is the ability of a powder to sink down into the water after it has been wetted. It is a functional property of powders and is affected by the particle density.
Dispersibility	This is the ability of a powder to be distributed throughout the water without the formation of lumps. It is a functional property of powders. The factors that affect this property are the reasonable absence of particles with a size > 250 μm or the absence of agglomerates.
Solubility	This is the dissolving rate or the total solubility. Absence of flecks and rapid swelling of particles are factors affecting the solubility.

The transformation of single particles into porous agglomerates is known as the instantizing process (Masters, 1991). The porous structure allows the quick entry of water with a subsequent sink, and the dispersion and dissolution of the solids. The agglomeration process changes the physical characteristics of the powder to such an extent that wetting, sinking, and dispersing are increased to a point where reconstitution occurs quickly.

Agglomerated powders are produced, among other practices, by integrating a fluid bed for product cooling to a spray dryer, as shown in Figure 6.13.

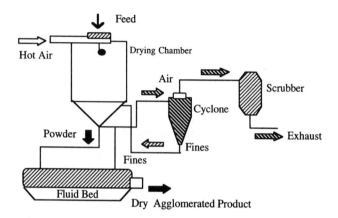

Figure 6.13.   Integrated fluid bed system into a spray dryer. (Adapted from Masters, 1991.)

6.7 FOAM SPRAY DRYING

Foam spray drying consists of the processing of liquid/ gas emulsions atomized into a drying chamber. The dispersion forms a foamed type particle when dry (Crosby and Weyl, 1977). The drying charateristic of foamed droplets are different than those of typical liquid droplets. Also, the properties and characteristics of foamed products are different than those of nonfoamed droplets such as reduced bulk density, increased particle size and porosity, difference in color, reduced moisture content, improved dispersability and dissolution, improved volatile retention, and minimized thermal degradation (Frey and King, 1986).

The foaming process may be performed by the direct mixing of a liquid and gas. During this operation, the size distribution of bubbles in the liquid is broad. A second technique consists of the supersaturation of the product to be dried with a gas. Bubble formation and growth are promoted by either a decrease in pressure or an increase in temperature. A third method consists of a desorption process wherein the gas is produced by a chemical reaction within the droplet.

The effect of pressure, P, on the volume fraction of insoluble gas in a foam is given by:

$$\alpha_{v2} = \frac{\alpha_{v1}\left(\dfrac{P_1}{P_2}\right)}{1 + \alpha_{v1}\left(\dfrac{P_1}{P_2} - 1\right)} \tag{43}$$

$$\alpha_v = \frac{v_v}{1 + v_v} \tag{44}$$

where v_v is the volume ratio of gas/feed, α_v is the volume fraction of gas in the foam, and P is the pressure. The effect of pressure on the final volume (V), surface area (A), and diameter (D) of foamed drops is expressed as:

$$1 + \alpha_{v1}\left(\frac{P_1}{P_2} - 1\right) = \frac{V_2}{V_1} = \left(\frac{A_2}{A_1}\right)^{3/2} = \left(\frac{D_2}{D_1}\right)^3 \tag{45}$$

The physical appearance of spray-dried foamed products may contain small uniformly dispersed voids, resemble thin- or thick-shelled hollow spheres, or it may contain large amount of small voids together. The effect of temperature on the volume fraction of insoluble gas, drop volumes, surface area, and diameter of the foamed drop can be expressed as follows:

$$\alpha_{v2} = \frac{\alpha_{v1}\left(\dfrac{T_1}{T_2}\right)}{1 + \alpha_{v1}\left(\dfrac{T_1}{T_2} - 1\right)} \tag{46}$$

$$1 + \alpha_{vl}\left(\frac{T_1}{T_2} - 1\right) = \frac{V_2}{V_1} = \left(\frac{A_2}{A_1}\right)^{3/2} = \left(\frac{D_2}{D_1}\right)^3 \tag{47}$$

Foaming did not suppress the temperatures experienced by the droplets during drying. The increased drying rate because of foam formation reduces the exposure time of the droplets to high temperature, thus reducing the thermal degradation of the product.

The drying of the droplets can be expressed as (Crank, 1956):

$$\frac{M_t}{M_\infty} = 1 - \frac{6}{\pi^2} \Sigma \frac{1}{n^2} \exp\left(\frac{-Dn^2\pi^2 t}{a^2}\right) \tag{48}$$

or

$$\frac{\left(M - M_e\right)}{\left(M_1 - M_e\right)} = \frac{6}{\pi^2} \Sigma \frac{1}{n^2} \exp\left(\frac{-D_{eff} n^2\pi^2 t}{r_1^2}\right) \tag{49}$$

where M_t is the moisture content at time t, M_∞ is the moisture content of the dried product, M is the average moisture content, M_e is the equilibrium moisture content, M_1 is the initial moisture content, n is the series index, D and D_{eff} are the diffusion coefficients, a is the sphere radius, and r_1 is the initial sphere radius.

The drying process of a foamed sphere has two limiting configurations: reduced sphere and spherical shell, as shown in Figure 6.14.

The diameters of the reduced sphere and the spherical shell can be expressed in terms of the initial diameter as follows:

$$\frac{D_{rs}}{D_1} = \left(1 - \alpha_v\right)^{0.33} \tag{50}$$

$$\frac{D_{is}}{D_1} = \left(1 - \alpha_v\right)^{0.33} \tag{51}$$

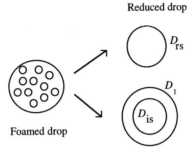

Figure 6.14. Limiting configurations of a foamed drop. (Adapted from Crosby and Weyl, 1977.)

Replacing Eq. (50) with Eq. (51), the reduced sphere model can be expressed as follows (Crosby and Weyl, 1977):

$$\frac{(M - M_e)}{M_1 - M_e} = \frac{6}{\pi^2} \Sigma \frac{1}{n^2} \exp\left(\frac{-4D_{eff} n^2 \pi^2 t}{D_{rs}^2}\right) \tag{52}$$

Meanwhile, the spherical shell model can be expressed as follows:

$$\frac{(M - M_e)}{(M_1 - M_e)} = \frac{3}{1 - \left(\dfrac{D_{is}}{D_i}\right)^3} \Sigma \frac{1}{(a_n^2 b_n)} \exp\left(\frac{-4a_n^2 D_{am} t}{D_1^2}\right) \tag{53}$$

$$\tan\left\{a_n\left[1 - \left(\frac{D_{is}}{D_i}\right)\right]\right\} = a_n\left(\frac{D_{is}}{D_i}\right) \tag{54}$$

$$b_n = \frac{\left(1 - \dfrac{D_{is}}{D_i}\right)}{2} - \sin\left\{\frac{2a_n\left[1 - \dfrac{D_{is}}{D_i}\right]}{4a_n}\right\} \tag{55}$$

Combining Eqs. (49), (50), and (52), the effective diffusivity for a foamed droplet can be expressed in terms of the diffusivity for a nonfoamed droplet:

$$D_{\text{eff}} = \frac{D_{\text{am}}}{(1 - \alpha_{\text{v}})^{2/3}} \qquad (56)$$

or using the spherical shell model:

$$D_{\text{eff}} = \frac{(D_{\text{am}} a_1^2)}{\pi^2} \qquad (57)$$

where D_{am} is the volume average diffusivity in a nonfoamed drop and a_n is the root of Eq. (54). Equations (56) and (57) demonstrate the effect of foaming on the drying process. The diffusivity increases as the fraction of gas increases, which results in an improved drying rate.

6.8 APPLICATIONS IN THE FOOD INDUSTRY

Several industrial applications have been developed utilizing spray drying. The most significant applications are related to the drying of milk, coffee, tea, eggs, whey proteins, enzymes, and microorganisms. A brief discussion and description of each is included in this section.

6.8.1 Milk Products

Skim milk powder with 4% moisture can be obtained from a concentrate of 45% to 55% solids feed using either a spray dryer chamber with external vibrated fluid beds as a cooler (Figure 6.15a) or a spray dryer with a pneumatic conveyor (Figure 6.15b).

Drying conditions where the exhaust air temperature is relatively low are used to produce milk with instant properties. The latter results in milk agglomerates that are

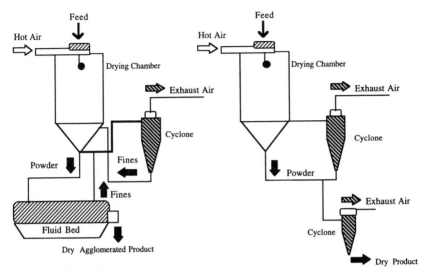

a. Spray dryer with external fluid-bed

b. Spray dryers with pneumatic conveyor

Figure 6.15.   Spray dryers used in dairy products. (Adapted from Masters, 1991.)

still high in moisture content when they are removed from the spray dryer. The final drying is performed on a fluid bed dryer. Fine particles recovered on the fluid bed are returned to the spray dryer for reagglomeration. Sometimes, the agglomeration is performed as a separate process, which results in a higher degree of agglomeration and very good instant properties (Masters, 1991).

Whole milk may be dried in systems such as those presented in Figure 6.15. The use of hammers at the chamber wall is necessary to prevent deposit formation caused by the fat content which makes the powder sticky. Milk concentrate of 40% to 50% solids is dried to 2% to 5% moisture content at 170 to 150°C inlet hot air. Figure 6.16 summarizes the steps in production of milk powder.

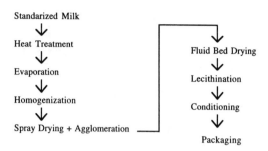

Figure 6.16.   Milk powder production steps. (Adapted from Masters, 1991.)

In addition to skim milk and whole milk, spray drying is used in the production of fat-filled milk. The product contains 26% to 28% fat and is prepared by adding vegetable or animal fat to whole milk before the drying operation occurs.

6.8.2 Instant Coffee

Extracted coffee of 15% to 30% solids can be concentrated to over 60% solids in a falling film evaporator or rotary thin film concentrator under vacuum, prior to spray drying (Upadhyaya and Kilara, 1986; Masters, 1991). The dried product consists of spherical particles about 300 μm and bulk density of 0.22 g/cm^3. Common drying parameters are 250°C on the inlet air and 110°C on the exhaust air (Upadhyaya and Kilara, 1986). Figure 6.17 summarizes a modern coffee plant layout.

6.8.3 Instant Tea

The processing of tea also requires the extraction of the material from leaves. The extraction leads to 5% to 20% solids which are concentrated up to 40% prior to spray drying using a falling film evaporator with an aroma

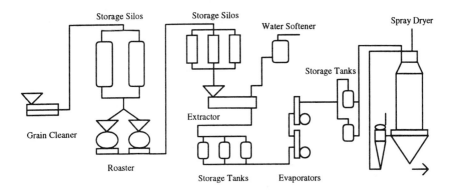

Figure 6.17.   Flowchart for a modern coffee plant. (Adapted from Masters, 1991.)

recovery system (Upadhyaya and Kilara, 1986; Masters, 1991). The use of carbon dioxide helps in the control of the bulk density of the final product. The drying is performed at 200 to 250°C with indirect heating (Masters, 1991). A co-current nozzle tower for tea spray drying is shown in Figure 6.18.

6.8.4 Dry Eggs

Egg products such as whole eggs, egg yolks, and egg whites are widely spray dried. The solid content of whole eggs is 25% to 27% and for egg yolk it is 45% to 48%. Prior to spray drying, the egg product is pasteurized at 64 to 66°C for 2 to 4 minutes to destroy *Salmonella* and any other undesirable microorganisms (Upadhyaya and Kilara, 1986). The drying operation is conducted using an inlet air temperature of 145 to 200°C in a co-current flow dryer with either a rotary or nozzle atomizer. It is recommended that the dried product be cooled down to 30°C or less as soon as possible to prevent heat damage (Upadhyaya and Kilara, 1986; Masters,

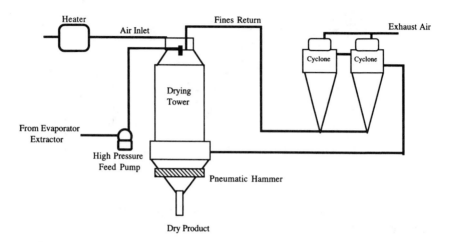

Figure 6.18.   Flowchart for a co-current nozzle spray dryer for tea. (Adapted from Masters, 1991.)

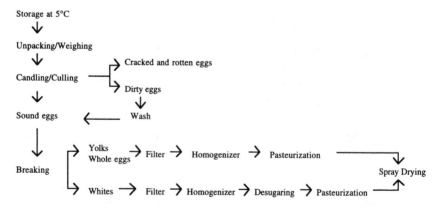

Figure 6.19.   Flowchart for spray dried egg products. (Adapted from Upadhyaya and Kilara, 1986.)

1991). Figure 6.19 illustrates a schematic process for producing dried egg products.

6.8.5 Spray Drying of Enzymes

Bulk enzymes are processed in standard co-current flow dryers. Process temperatures are: inlet air 143°C, outlet air 71°C, product outlet temperature 55°C, and product moisture content 10 to 20%. Two-stage drying avoids the loss of enzymatic activity. Table 6.5 presents some spray-dried enzymes that are commercially available as well as their uses.

6.8.6 Microorganisms and Yeasts

The spray drying of bacteria can be carried out using nitrogen as a drying medium. Feed solid content of 6%, inlet temperature under 95°C, product recovery temperature of 50°C, and a final moisture content of 4% are examples of process parameters. Some spray-dried bacteria are *Escherichia coli, Lactobacillus casei, Streptococcus lac-*

Table 6.5.   Spray dried enzymes.

Enzyme	Use
Amylase	Baking, textiles, brewing
Protease	Brewing, tenderizing, detergents, cheesemaking, tanning
Glucose oxidase	Beverages
Pectinase	Fermentation, juice clarification
Lipase	Detergents
Trypsin	Wound debridement
Rennin	Cheesemaking
Lactase	Ice cream
Cellulase	Cellulose breakdown

From Masters, 1991.

tis, Bacillus subtilis, and *Serratia marcescens* (Masters, 1991).

Johnson and Etzel (1993) reported the preparation of a spray-dried starter culture from lactic acid bacteria. This culture was stable and nonviable while preserving the intracellular peptidase and had no increase in lactic acid production when used in cheese. The process conditions for spray drying range from high solids feed solution with a high outlet air temperature for starter cultures used in the manufacture of low-fat cheese, to a low solids solution and low outlet air temperature for normal starter culture. The pretreatment of the microorganisms (freezing, adding dextrin) plays an important role in the viability of the spray-dried cells, as well as the final moisture content and storage conditions.

Yeasts contain proteins, carbohydrates, lipids, minerals, and vitamins. *Turolopsis utilis* strains are the most

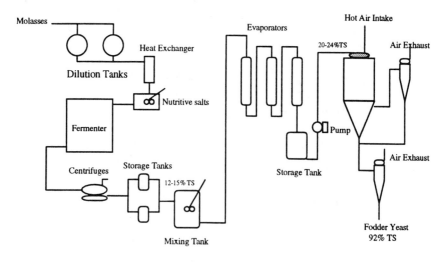

Figure 6.20.   Schematic of a fodder yeast spray drying plant. (Adapted from Masters, 1991.)

important in the manufacture of food yeasts (Masters, 1991). The yeast is grown and propagated in sulfite waste liquor, molasses, whey, maize, or sugar-containing by-products. The term *fodder yeast* applies when the yeast is propagated solely for animal feeding. Figure 6.20 illustrates a flow diagram intended for this type of product. The concentration of yeast solids in the fermentation of liquor range from 4% to 8% before centrifugation, and 12 to 15% total solids after centrifugation. A total of 22% solids can be obtained by evaporation. The spray drying temperature is 300 to 350°C for inlet air and 100°C for exhaust air.

A similar drying process is used on *Saccharomyces cerevisiae.* The feed will contain about 50% solids prior to spray drying. A cool, low-humidity environment is required to handle the final product because of hygroscopicity.

6.8.7 Spray Drying of Whey Proteins

Gruetzmacher and Bradley (1991) described the production of spray-dried coffee whitener using acid whey. The use of acid whey as coffee whitener is an application that can solve a major problem in the dairy industry concerning the disposal of the raw material and its impact on the environment. Spray-dried coffee whiteners have the advantage of lower cost, convenience, ease of handling, and improved shelf life without refrigeration. Coffee whiteners are used as substitute for cream, evaporated milk, and fresh milk in coffee, tea, and cocoa. Figure 6.21 summarizes the flow diagram proposed by Gruetzmacher and Bradley (1991). The whey is heated and mixed with sodium stearoyl lactylate, titanium dioxide, corn syrup, polysorbate 60, dipotassium phosphate, lecithin, colors, and flavors in a temperature range of 55 to 65°C. The mixture is homogenized prior to spray drying. The feed to the

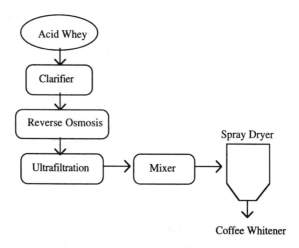

Figure 6.21. Flowchart for the production of spray-dried coffee whitener from acid whey (Gruetzmacher and Bradley, 1991).

spray dryer has 65% total solids and is dried to 3% moisture content (Gruetzmacher and Bradley, 1991)

6.9 CONCLUDING REMARKS

Spray drying is an important processing operation in the food industry. The drying characteristics of any food intended for spray drying need to be evaluated on a pilot plant scale prior to an industrial application, but the final operation conditions should be defined only for the spray dryer designed or selected for the application. The most important characteristic of a spray dried product is the uniformity of the final product. The instantizing properties of the spray-dried product are based on the post-drying processing, which leads to an agglomerated product.

6.10 NOMENCLATURE

α	Angle of liquid release or thermal diffusivity
β	Factor used in Eq. (17)
ρ	Density (w, droplet; a, air; g, gas), kg/m3
η	Thermal efficiency
ν	Kinematic viscosity
μ	Viscosity, cP
σ	Surface tension, dynes/cm
A	Area, m^2
a	Constant
b	Liquid film thickness
C_d	Drag coefficient
C_p	Heat capacity, kJ/kg K
c_p	Specific heat of droplet
Cv	Velocity coefficient
D	Diameter (wet, wet droplet; dry, dry particle; vs, Sauter diameter; wheeldiameter) or Diffusivities (eff, effective; am, volume average)
d	Constant Eq. (16)
$\boldsymbol{d}$	Pure liquid droplet size, m
E_h	Kinetic energy
F_s	Solid content, kg
G	Dry air flow, kg/s
$\mathbf{g}$	Gravitational constant, 9.81 m/s2
H	Humidity, kg of water/kg of dry air
h	Vane height (m), pressure head (kPa), or heat transfer coefficient (kJ/m2K)
K	Constant
k_y	Mass transfer coefficient, kg/s m^2
k	Thermal diffusivity
l	Length, m
N	Wheel velocity as rpm
Nu	Nusselt number
n	Number of vanes in wheel or constant in Eq. (7)

Pr   Prandlt number
P_k Power, kW h/ton feed
DP   Pressure drop, kPa
p   Constant
Q   Flow rate, kg/h
q   Enthalpy (s, solids; a, air), kJ/kg or a constant
Re   Reynolds number
r   Radius, (1, inlet; 2, orifice), m or a constant
Sc   Schmidt number
Sh   Sherwood number
s   Constant
U   Component of velocity (r, radial; t, tangential; h,horizontal; v, vertical;rel, relative), m/s
V   Velocity, m/s
w   Moisture content, kg of water/kg of dry solids

6.11 REFERENCES

Charlesworth, D. H. and Marshall, W. R. 1960. Evaporation from drops containing dissolved solids. *AIChE J.* 6(1):9–23.

Charm, S. E. 1978. Dehydration of foods. In *The Fundamentals of Food Engineering*, Third edition, AVI Publishing, Westport, CT.

Crank, J. 1956. *The Mathematics of Diffusion*. Oxford University Press, London, UK.

Crosby, E. J. and Weyl, R. W. 1977. Foam spray drying: general principles. *Chem. Eng. Symp. Series* 73. 163:82–94.

Crowe, C. T. 1980. Modeling spray-air contact in spray drying systems. In *Advances in Drying*, Vol. 1, edited by A. S. Mujumdar. Hemisphere Publishing, New York.

Dittman, F. W. and Cook, E. M. 1977. Establishing the parameters for a spray dryer. *Chem. Eng.* 84(2):108–112.

Frey, D. D. and King, C. J. 1986. Experimental and theoretical investigation of foam spray drying. 2. Experimental investigation of volatiles loss during foam-spray drying. *Ind. Eng. Chem. Fundam.* 25:730–735.

Gruetzmacher, T. J. and Bradley, R. L. 1991. Acid whey as a replacement for sodium caseinate in spray dried coffee whiteners. *J. Dairy Sci.* 74:2838–2849.

Heldman, D. R. and Singh, R. P. 1981. Food dehydration. In *Food Process Engineering*, Second edition, AVI Publishing, Westport, CT.

Johnson, J. A. C. and Etzel, M. R. 1993. Inactivation of lactic acid bacteria during spray drying. *AIChE Symp. Series* 89(297):98–107.

King, C. J., Kieckbusch, T. G., and Greenwald, C. G. 1984. Food quality factors in spray drying. In *Advances in Drying*, Vol. 3, edited by A. S. Mujumdar. Hemisphere Publishing, New York.

Marshall, W. R. 1954. *Atomization and spray drying.* Chemical Eng. Process Monogr. Series. 50(2). New York.

Masters, K. 1991. *Spray Drying Handbook*, Fifth edition. Longman Group Limited, UK.

Pisecky, J. 1986. Standards, specifications, and test methods for dry milk products. In *Concentration and Drying of Foods*, edited by D. MacCarthy. Elsevier Applied Science, New York.

Shaw, F. V. 1994. Fresh options in drying. *Chem. Eng.* 101(7):76–84.

Upadhyaya, R. L. and Kilara, A. 1986. *Drying heat sensitive products. In Drying of Solids. Recent International Developments,* edited by A. S. Mujumdar. John Wiley & Sons, New York.

CHAPTER 7

FREEZE DEHYDRATION

7.0 INTRODUCTION

Freeze drying was developed to overcome the loss of the compounds responsible for flavor and aroma in foods, which are lost during conventional drying operations (Karel, 1975; Dalgleish, 1990). The freeze drying process consists mainly of two steps: (1) the product is frozen, and (2) the product is dried by direct sublimation of the ice under reduced pressure. Freeze drying, or lyophilization, was initially introduced in the 1940s on a large scale for the production of dry plasma and blood products (Rey, 1975). Later, antibiotics and biological materials were prepared on an industrial scale by freeze drying. Figure 7.1 shows a basic configuration of a freeze drying system.

Freeze drying has been shown to be an attractive method for extending the shelf life of foods (Ma and Arsem, 1982). The drying of food products in freeze drying has two main characteristics (Longmore, 1971):

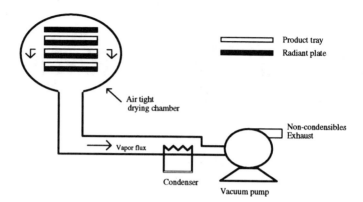

Figure 7.1.   Basic freeze drying system. (Adapted from Liapis and Marchello, 1984).

1. Virtual absence of air during processing: The absence of air and low temperature prevent deterioration due to oxidation or chemical modification of the product.

2. Drying at temperatures lower than ambient temperature: Products that decompose or undergo changes in structure, texture, appearance, and/or flavor as a consequence of high temperature can be dried under vacuum with minimum damage.

Freeze-dried products that are properly packaged can be stored for an unlimited period of time, while retaining most of the physical, chemical, biological, and organoleptic properties of their fresh state. Freeze drying reduces loss of quality due to deleterious chemical reactions caused by enzymatic and nonenzymatic browning. However, the oxidation of lipids, induced by the low moisture levels achieved during drying, is a major concern for freeze-dried products. Lipid oxidation reactions are controlled by packaging in oxygen-impermeable containers. Nonenzymatic browning is avoided because of the rapid transition from high to low moisture content during the process. The use of low temperature ranges also avoids protein denaturation in freeze-dried products (Okos et al., 1992).

Freeze-dried products can be reconstituted to their original shape and structure by the addition of liquid. The spongelike structure of the dried product allows a rapid rehydration process. The rehydrated product characteristics are similar to those in a fresh product. The porosity of freeze-dried products allows for more complete and rapid rehydration than is possible with air-dried products.

The major disadvantages of freeze drying are the energy cost and the drying time (Liapis and Marchello, 1984). It is important to note that in the context of this book freeze drying will be discussed in terms of the water-food relationship only, but the operation can also be used to remove other types of liquids in complex mixtures. The latter is outside of the scope of this work.

Some food products that are commercially freeze-dried include extracts (coffee and tea), vegetables, fruits, meats, and fish (Schwartzberg, 1982; Dalgleish, 1990). Freeze-dried products are lightweight (10% to 15% original weight) and do not require refrigeration. Moisture levels as low as 2% are reached with freeze drying. Steaks, fish, and chicken can be dried without crushing or shredding the product (Sacharow and Griffin, 1970).

7.1 FUNDAMENTALS OF FREEZE DEHYDRATION

The process consists of two main stages: freezing and drying. Freezing must be very rapid to obtain a product with small ice crystals and in an amorphous state (Mellor, 1978). The drying process involves lowering the pressure to allow ice sublimation. Figure 7.2 presents the phase diagram of water, and Figure 7.3 presents the freeze drying steps.

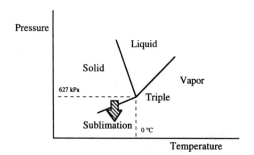

Figure 7.2.   Phase diagram of water. (Adapted from Karel, 1975.)

Three important design variables to be considered in freeze drying are: (1) vacuum inside the chamber, (2) radiant energy flux applied to the food, and (3) the temperature of the condenser. The initial drying rate is high because there is little resistance to either heat or mass flux. However, a buildup of a resistive layer around the frozen material slows down the rate as the drying proceeds. The dry layer around the product serves as an insulation material that affects the heat transfer to the ice front. Also, the mass transfer from the ice front is reduced

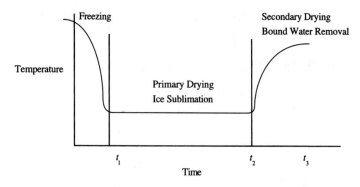

Figure 7.3.   Freeze drying steps (Adapted from Mellor, 1978.)

as the thickness of the dry layer is increased. This is because of a reduction in the diffusion process from the sublimation interface to the product surface.

7.1.1 Freezing Step

The freezing temperature and time on foodstuffs is a function of the solutes in solution (Schwartzberg, 1982). Pure water freezing temperature remains constant at the freezing point until all the water is frozen. In the case of food, the freezing temperature is lower than that of pure water. Because the solutes become more concentrated in the unfrozen portion of the mix, the freezing point temperature continually decreases until all the solution is frozen. At the end of the freezing process, the entire mass should become rigid, forming an eutectic consisting of ice crystals and food components (Mellor, 1978). The eutectic state is required to ensure the removal of the water by sublimation only and not by a combination of both sublimation and evaporation. Melting and inadequate freezing should be avoided because the formation of frothy and gummy substances will appear in the final product.

The permeability of the frozen surface layer can be affected by the migration of soluble components during the freezing step. However, the removal of a thin surface layer of the frozen product, or freezing under conditions inhibiting the separation of the concentrate phase, result in better drying rates (Karel, 1975).

7.1.2 Drying Steps—Primary and Secondary Drying

Two drying steps can be identified during lyophilization (King, 1970; Mellor, 1978). The primary drying step involves sublimation of ice under vacuum. The ice sublimes when the energy for the latent heat is supplied.

Because of the low pressure, the water vapor generated in the sublimation interface is removed through the outer porous layers of the product as shown in Figure 7.4a. The condenser prevents the return of the water vapor to the product. The driving force for the sublimation is essentially the difference in pressure between the water vapor pressure at the ice interphase and the partial pressure of water vapor in the drying chamber. The energy for sublimation (latent heat) can be supplied by radiation and conduction through the frozen product, or by irradiating the water molecules with microwaves (Mellor, 1978; Arsem and Ma, 1990).

The secondary step (Figure 7.4b) begins when no more ice (from unbound water) is in the product and the moisture comes from partially bound water in the drying material. At this time the heating rate must decrease to maintain the temperature of the product below 30 to 50°C which will prevent collapse of the material (Mellor, 1978). If the solid matrix becomes too hot, the structure collapses, which in turn decreases the sublimation rate from

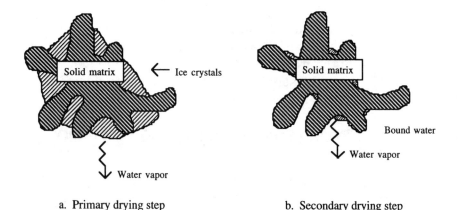

a. Primary drying step   b. Secondary drying step

Figure 7.4.   Removal of water during freeze drying.

the ice front within the product as discussed by Bellows and King (1973). The secondary drying step will take up to a third of the total drying cycle to desorb the moisture from the internal surface within the dried product.

7.2 COMBINED HEAT AND MASS TRANSFER

The mass and heat transfer phenomena during freeze drying can be summarized in terms of the diffusion of water vapor from the sublimation front, and the radiation and conduction of heat from the heating platen.

7.2.1 Steady-State Model

The energy required to sustain the sublimation is assumed to be equal to the radiant or conduction flux provided by the temperature gradient between frozen product and the heat source in the drying chamber (King, 1970; Schwartzberg, 1982; Okos et al., 1992). The water sublimes below the triple point under pressures of 627 Pa or less (Okos, et al., 1992). The sublimation interface is located immediately above the ice front, and water removal occurs at or very close to the sublimation interface. Figures 7.5a and 7.5b show the heat and mass flow during drying on frozen slabs.

The heat flux due to convection and conduction at the sublimation surface (Figure 7.5a) can be expressed as (Shukla, 1990; Okos et al., 1992):

$$q = h\left(T_e - T_s\right) = k \frac{\left(T_s - T_f\right)}{\left(L_2 - L_1\right)} \tag{1}$$

where q is the heat flux (W/m^2), h is the external heat transfer coefficient (W/K $\cdot$ m^2), T_e is the external temperature of the vaporizing gas ($^\circ$C), T_s is the surface temperature of the dry solid ($^\circ$C), k is the thermal conductivity of

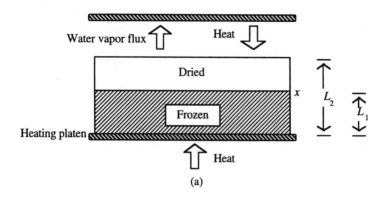

(a)

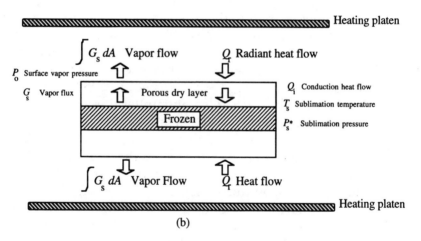

(b)

Figure 7.5.   Mass and heat flow during freeze drying: (a) single side drying; (b) symmetrical arrangement (Adapted from Schwartzberg, 1982.)

the dry solid (W/m K), $(L_2 - L_1)$ is the thickness of the dry layer (m), and T_f is the temperature of the sublimation front or ice front. Table 7.1 shows the thermal conductivity values and sublimation temperatures reported for some freeze-dried food products.

Table 7.1.   Thermal conductivity and sublimation temperature of freeze-dried food products.

Product	k (W/m K)	Product	T_{sub} (°C)
Beef	0.035–0.038	Chicken	−21
Mushrooms	0.010	Shrimp	−18
Peach	0.016	Salmon	−29
Apple	0.016–0.035	Beef	−14
Coffee extract—25%	0.033	Whole egg	−17
Milk	0.022	Apple	− 7
Gelatin	0.016	Coffee	−23
Turkey	0.014	Carrots	−25

From Schwartzberg (1982).

The water vapor flux from the sublimation front is given by (King, 1970; Okos et al., 1992):

$$N_a = \frac{D'\left(P_{fw} - P_{sw}\right)}{RT\left(L_2 - L_1\right)} = K_g\left(P_{sw} - P_{ew}\right) \tag{2}$$

where N_a is the water vapor flux (kgmol/s m^2), D' is the average effective diffusivity of water vapor in the dry layer (m^2/s), R is the universal gas constant, T is the average temperature in the dry layer (°C), P_{fw} is the partial pressure of water vapor in equilibrium with the sublimation ice front (atm), P_{sw} is the partial pressure of water vapor at the surface (atm), K_g is the external mass transfer coefficient (kg-mol/s m^2 atm), and P_{ew} is the partial pressure of water vapor in the external bulk gas phase (atm).

Equations (1) and (2) can be rearranged to express q and N_a in terms of the external operating conditions as follows (King, 1970; Okos et al., 1992):

$$q = \frac{T_e - T_f}{\left(\dfrac{1}{h} + \dfrac{\left(L_2 - L_1\right)}{k}\right)} \tag{3}$$

and

$$N_a = \frac{\left(P_{fw} - P_{ew}\right)}{\left(\dfrac{1}{K_g} + \dfrac{RT\left(L_2 - L_1\right)}{D'}\right)} \tag{4}$$

The constants h and K_g depend on both the gas velocities and the dryer while k and D' depend on the nature of the dried material. T_e and P_{ew} are set by the external operating conditions. Equations (3) and (4) can be related through the latent heat of sublimation of ice (ΔH_s, J/kg-mol) as follows:

$$q = \Delta H_s N_a \tag{5}$$

Combining Eqs. (1), (4), and (5) gives:

$$\frac{k\left(T_s - T_f\right)}{\left(L_2 - L_1\right)} = \frac{\Delta H_s\left(P_{fw} - P_{ew}\right)}{\left(\dfrac{1}{K_g} + \dfrac{RT\left(L_2 - L_1\right)}{D'}\right)} \tag{6}$$

or

$$h\left(T_e - T_s\right) = \frac{\Delta H_s\left(P_{fw} - P_{ew}\right)}{\left(\dfrac{1}{K_g} + \dfrac{RT\left(L_2 - L_1\right)}{D'}\right)} \tag{7}$$

An increase in T_e or T_s causes an increase in the drying rate as expressed by Eqs. (6) and (7). T_s is limited by the heat sensitivity of the material, and T_f must be below the collapse temperature of the product. The sensitivity is defined in terms of degradation reactions, while the collapse temperature is defined in terms of the deformation of the porous structure of the dried layer.

The rate of freeze drying can be expressed as:

$$N_a = \left(\frac{L}{2M_aV_s}\right)\frac{-dz}{dt} \tag{8}$$

where L is the total thickness of the solid, z is the thickness of the dried layer, t is time, M_a is the molecular weight of water, V_s is the volume of the solid occupied by a unit kg of water, expressed as $V_s = 1\,/(\,X_0\rho_s)$, X_0 is the initial moisture content, and ρ_s is the bulk density of the dry solid. A similar approach was considered by Schwartzberg (1982) to describe the freeze drying process considering heat and mass transfer through two of the product faces as shown in Figure 7.5b. The water vapor flux in the system described by Schwartzberg can be expressed as:

$$G_s = \frac{K_p\left(P_s^* - P_0\right)}{z} \tag{9}$$

$$= \frac{K_p\left(P_s^* - P_c^*\right)}{z} \tag{10}$$

$$= \rho\frac{\left(X_0 - X_f\right)}{\left(1 + X_0\right)}\frac{dz}{dt} \tag{11}$$

where P_s^* is the vapor pressure of water at the sublimation interface, K_p is the permeability of the dry layer, ρ is the density of the frozen portion of the slab, X_0 is the initial moisture content of the food (mass water/mass dry solids), X_f is the final moisture content, z is the thickness of the dried layer, P_0 is the partial pressure of the water at the surface, P_c^* is the condenser pressure that will be equal to P_0 unless noncondensables are introduced into the dryer (Schwartzberg, 1982), and t is the drying time. The first part of Eq. (11) represents the change in water content per unit of volume of frozen product. The change in thickness of the dried layer change is a function of time, given that the total surface area remains constant. Table 7.2 presents some permeability values reported for freeze-dried foods. Equation (11) can be integrated and the drying time expressed as:

$$t_s = \rho \frac{(X_o - X_f)}{2K_p(1+X_o)} \frac{a^2}{(P_s^* - P_o)} \tag{12}$$

where a is the half-thickness of the slab.

The heat required for sublimation is assumed to be equal to the radiant heat flux Q_r and can be expressed as:

$$Q_r = Q_i = G_s H_s \tag{13}$$

where Q_i is the internal heat flux inside the slab and H_s is the average latent heat of vapor (Schwartzberg, 1982). The internal heat flux, Q_i, can be expressed in terms of the dry layer characteristics as follows:

$$Q_i = \frac{K_t(T_o - T_s)}{z} \tag{14}$$

where K_t is the thermal conductivity of the dry layer, T_o is the surface temperature of the slab, and T_s is the sublimation temperature. Substituting Eq. (11) for G_s in Eq. (14):

$$\frac{K_t(T_o - T_s)}{z} = \rho \frac{(X_o - X_f)}{1+X_o} H_s \frac{dz}{dt} \tag{15}$$

Assuming that T_o, T_s and K_t remain constant:

$$t_s = \rho \frac{(X_o - X_f)}{1+X_o} H_s \frac{a^2}{K_t(T_o - T_s)} \tag{16}$$

By combining the heat-transfer and mass-transfer relations for the drying time, t_s, the following relationship is

Table 7.2. Permeability of freeze-dried foods.

Product	Permeability (10^{-9} kg/m s μmHg)
Whole milk	2.7–5.3
Apples	3.3–6.0
Carrot	2.0–5.6
Potato	1.3
Banana	1.1
Beef	0.7–4.4
Coffee 20% solids	4.0–8.6
Coffee 30% solids	3.0
Tomato 22°Brix	2.1
Fish	8.7

From Schwartzberg (1982).

obtained which is a function of the dry layer properties and the operation conditions during freeze drying:

$$\left(P_s^* - P_o\right) = \left(\frac{-K_t}{K_p H_s}\right)\left(T_s - T_o\right) \tag{17}$$

Equation (17) describes a straight line if P_s^* and T_s are considered the variables, T_o and P_o are fixed and K_t and K_p are independent of P_s^* and T_s (Schwartzberg, 1982). Values for P^* can be calculated from the sublimation equation as follows:

$$\ln P^* = 30.9526 - \frac{6153.1}{T} \tag{18}$$

where T is the absolute temperature in degrees Kelvin (K). Equations (17) and (18) can be used together to define the final operation condition during a freeze drying process.

Another important variable is the surface temperature, T_o, which is controlled by the rate of heat transfer from the heating platen:

$$Q_r = F_{op}\sigma[T_p^4 - T_0^4] \tag{19}$$

where σ is the thermal diffusivity and F_{op} is the exchange factor defined as:

$$F_{op} = \frac{1}{\left(\dfrac{1}{\epsilon_0} + \dfrac{1}{\epsilon_p} - 1\right)} \tag{20}$$

where ϵ_0 is the emissivity of the load surface and ϵ_p is that of the platen. Assuming $\epsilon_p \approx 1.0$, $F_{op} \approx \epsilon_0$ (Schwartzberg, 1982) and combining Eqs. (14), (16), and (19), T_p can be expressed as follows:

$$T_p^4 = T_0^4 + \frac{K_t\left(T_0 - T_s\right)}{\left(\dfrac{2K_t\left(T_0 - T_s\right)\left(1 + X_0\right)}{H_s\rho\left(X_0 - X_f\right)} t + z^2\right)^{0.5}} \tag{21}$$

EXAMPLE 1

A 15 mm thick slab of beef is subjected to freeze drying. The piece of meat has an initial moisture content of 80%. The final moisture content desired is 5%. The initial density of the beef is 1067 kg/m³. The sublimation pressure is maintained at 250 μmHg and the condenser pressure is maintained at 100 μmHg. $K_p = 0.7E\text{-}9$ kg/m s μmHg (Schwartzberg, 1982). Estimate the drying time.

Answer

$$a = 0.0075 \text{ m} \qquad X_0 = 0.8/0.2 = 4$$

$$X_f = 0.05/0.95 = 0.0526 \qquad P_0 = P_c^*$$

Using Equation (12)

$$t_s = \rho \frac{\left(X_0 - X_f\right)}{2K_p\left(1 + X_0\right)} \frac{a^2}{\left(P_s^* - P_0\right)}$$

$$= \frac{1067(4 - 0.0526)(0.0075)^2}{2((0.7E - 9)(1 + 4)(250 - 100))}$$

$$= 226E - 3 \text{ s or } 62.8 \text{ h.}$$

EXAMPLE 2

The freeze drying of a beef slab is done while maintaining a surface temperature of 35°C and a condenser temperature of −30°C. K_p for the beef is 0.7E-9 kg/m s μmHg and K_t = 0.035 W/m K (Schwartzberg, 1982). Estimate T_s and P_s.

Answer

The value of P^* is obtained from the equilibrium relationship described by Eq. (18); T_c = −30°C, P^* = 285 μmHg. Assuming that $P_c^* = P_o$, P_o = 285 μmHg. Equation (17) is used to estimate the slope of the line which describe the operating conditions for K_t = 0.035 W/m K and K_p = 0.7 E-9 kg/m s μmHg.

$$-K_t/K_pH_s = -17.55$$

The graphical solution is presented in Figure 7.6.

7.2.2 Non-Steady-State Model

Liapis and Marchello (1984) discussed a sorption– sublimation model that describes freeze drying based on an unsteady state of energy and mass balance in the dried and frozen regions.

Heat balance—dry layer

$$\frac{\partial T_D}{\partial t} = \alpha_D \frac{\partial^2 T_D}{\partial x^2} - \frac{N_t C_{pg}}{\rho_{De} C_{pDe}} \frac{\partial T_D}{\partial x} + \frac{\Delta H_s \rho_D}{\rho_{De} C_{pDe}} \frac{\partial X}{\partial t} \qquad (22)$$

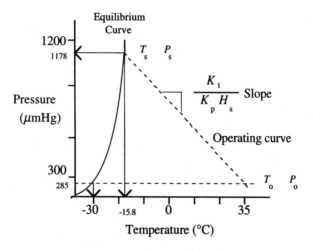

Figure 7.6. Graphical solution of Example 2.

Heat balance—frozen layer

$$\frac{\partial T_F}{\partial t} = \alpha_F \frac{\partial^2 T_F}{\partial x^2} \tag{23}$$

where C_{pDe} is the effective heat capacity of the dried layer, C_{pg} is the heat capacity of the gas phase, α is the thermal diffusivity for the dried (D) or frozen (F) layers, T is the temperature, and N_t is the total flux (water vapor and inert gas). The continuity equations for the dried region as, defined by Liapis and Marchello (1984), are given by:

$$\frac{1}{R}\frac{\partial\left(\dfrac{P_w}{T}\right)}{\partial t} = \left(\frac{-1}{M_w \epsilon}\right)\frac{\partial N_w}{\partial x} - \left(\frac{\rho_{De}}{M_w \epsilon}\right)\frac{\partial X}{\partial t} \tag{24}$$

$$\frac{1}{R}\frac{\partial\left(\dfrac{P_{in}}{T}\right)}{\partial t} = \left(\frac{-1}{M_{in}\epsilon}\right)\frac{\partial N_{in}}{\partial x} \tag{25}$$

$$\frac{\partial X}{\partial t} = K_{g}\left(X^{*} - X\right) \tag{26}$$

where X^* is the weight fraction of water in the solid in local equilibrium with the partial pressure of water vapor obtained from sorption isotherm data, ϵ is the void fraction, M_{in} is the molecular weight of the inert gas contained in the frozen layer, M_w is the molecular weight of water, N_w is the water flux, N_{in} is the inert gas flux, and P_{in} is the partial pressure of the inert gas.

The diffusion equations can be expressed as:

$$N_{w} = \left(\frac{-M_{w}}{RT}\right)\left(k_{1}\frac{\partial P_{w}}{\partial x} + k_{2}P_{w}\left(\frac{\partial P_{x}}{\partial x} + \frac{\partial P_{in}}{\partial x}\right)\right) \tag{27}$$

$$N_{in} = \left(\frac{-M_{in}}{RT}\right)\left(k_{3}\frac{\partial P_{in}}{\partial x} + k_{4}P_{in}\left(\frac{\partial P_{in}}{\partial x} + \frac{\partial P_{w}}{\partial x}\right)\right) \tag{28}$$

where k_1, k_2, k_3, and k_4 are the diffusivity coefficients and P_x is the partial pressure of water at the frozen interface.

7.3 STRUCTURAL CHANGES AND VOLATILE RETENTION DURING FREEZE DRYING

The freezing step causes separation of the aqueous solutions present in foods into a two-phase mixture: ice and a concentrated solution of solutes. As a consequence of the ice formation, the shrinkage of the product is reduced and the dried product attains a spongelike structure that promotes an easy rehydration. Also, the mobility of the concentrate phase is so low that no structural changes occur during drying, which contributes to the spongelike structure mentioned previously (Van Arsdel and Copley, 1963; Massaldi and King, 1974; Karel, 1975; Bruin and Luyben, 1980). Another characterstic of the product is the low bulk density structure of freeze-dried food products resulting from the needlelike void spaces that were previously occupied by the ice crystals.

The collapse phenomenon during freeze drying occurs when the mobility of the concentrate phase increases. A change in the bulk density is obtained when the product is heated to a certain temperature. This change is known as collapse (Bellows and King, 1973; Karel and Flink, 1983). The phenomenon is noticeable as a shrinkage of the dry cake and it is attributed to a reduction in the viscosity of the solid matrix as presented in Figure 7.7. The temperature at which collapse occurs is a function of the moisture content and the solutes in the food. Intermediate locations on the drying layer are more susceptible to collapse rather than the outermost layers. Table 7.3 shows the collapse temperature for some food products.

The volatiles in freeze-dried products are retained by entrapment within the dry food matrix (Karel, 1975; Karel and Flink, 1983). The concept of microregion permeabil-

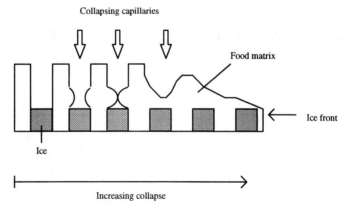

Figure 7.7. Progressive stages of collapse. (Adapted from Bellows and King, 1973.)

Table 7.3.   Collapse temperature of food products.

Product	Temperature (°C)
Potato	−1.5
Fructose	−48
Glucose	−40
Sucrose	−22 to −32
Lactose	−19 to −31
Gelatin	−8
Orange juice	−24
Grapefruit juice	−30.5
Lemon juice	−36.5
Coffee extract	−20 to −26

From Bellows and King (1973) and Schwartzberg (1982).

ity discussed by Karel (1975) and Bruin and Luyben (1980) is based on the premise that adsorption of volatiles in the dry layer is not responsible for their retention. The microregions are small areas in the dry material where the volatiles are localized and strongly bound (Karel, 1975; Karel and Flink, 1983).

The sorption of water causes an increase in microregion permeability and a subsequent loss of volatiles. The carbohydrate–carbohydrate bonds are replaced by carbohydrate–water bonds, increasing the mobility of the carbohydrates and increasing the permeability to volatiles. However, at the critical moisture content, X_c, the microregions are sealed and no further loss of volatiles occurs.

7.4 FREEZE-DEHYDRATION RELATED PROCESSES

Prefreezing, preconcentration, condensing, and defrosting are processes related to freeze drying. Different methods are used to perform the freezing step: individual quick freezing, fluidized bed freezing, immersion freezing, plate freezing, air blast freezing, and the use of scraped surface heat exchangers.

Cost and efficiency are always factors to be considered when designing a freeze drying process. The preconcentration of the product prior to freezing and drying is an important consideration in the reduction of the drying time and the increase in process efficiency.

During the drying process it is very important to remove the water vapor released from the product. In freeze drying, the water vapor is sublimated against a surface condenser cooled down to –60°C. This is done to prevent the water from returning to the product or from passing into the vacuum pumps. As the drying progresses, the amount of ice deposited in the condenser grows at a rate proportional to the sublimation rate of the product. The removal of the ice is called defrosting and is necessary to maintain a proper rate of heat transfer at the condenser.

7.4.1 Prefreezing

Prefreezing is used to reduce the freeze drying time cycle. Once the product is frozen the freezing conditions are maintained by a refrigeration system until the product is ready for drying. Freezing systems are classified into two groups: direct contact and indirect contact systems (Heldman, 1992).

7.4.1.1 Direct Contact

In direct contact refrigeration systems the refrigerant is in direct contact with the product surface. The freezing system attempts to bring the refrigerant into contact with as many surfaces as possible, as shown in Figure 7.8 (Heldman, 1992).

The refrigerant types used for these systems include low-temperature air and selected liquid. Some of the designs used in direct contact systems include: *individual quick freezing* (IQF), in which individual small pieces of product are exposed directly to low-temperature air;

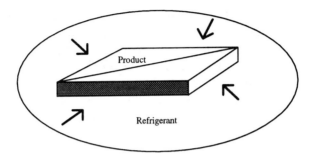

Figure 7.8. Schematic diagram of direct contact freezing. (Adapted from Heldman, 1992.)

fluidized bed freezing, which is a modified version of an IQF system in which the product pieces are maintained in a fluidized state, which creates very high convective heat transfer coefficients; and *immersion freezing*, in which the product is exposed to a liquid refrigerant that is undergoing a phase change as the freezing occurs.

7.4.1.2 Indirect Contact

In this type of system, the food is separated from the refrigerant by some type of barrier as shown in Figure 7.9.

Some designs used in indirect contact systems are: *plate freezing*, in which the product is maintained between plates during the freezing process; *air blast freez-*

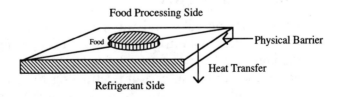

Figure 7.9. Indirect contact freezing system. (Adapted from Heldman, 1992.)

ing, in which the product package represents the barrier between the product and the cold air; and *scraped surface heat exchangers*, in which the product passes by one side of a heat exchanger while the refrigerant passes by the other. A rotor is used to provide movement to the frozen product at the heat exchanger surface. The use of this type of system results in the removal of 60% to 80% of the fusion latent heat from a liquid food (Heldman, 1992).

7.4.2 Preconcentration

The removal of water from foods before submitting them to a formal drying process is known as preconcentration. Water may be removed in the solid form of ice, in the liquid form by reverse osmosis, or as a gas in the form of vapor by evaporation (Kessler, 1986). In the case of freeze drying, freeze concentration is the most compatible operation because the low processing temperature make it suitable for the concentration of thermally sensitive food products. Concentration levels of 40% to 55% by weight can be attained. A general scheme of a freeze concentration process is shown in Figure 7.10.

The separation process of ice crystals is done in a *wash column*. This type of operation is used almost exclusively in freeze concentration systems for food products (Hartel, 1992).

7.4.3 Condensation

Freeze drying takes advantage of surface condensers, which allow a uniform contact of cooling medium through a heat transfer surface (Hartel, 1992). This prevents the water vapor from returning to the product or passing into the vacuum pumps. Different types of condensers are available: shell and tube, spiral plate, spiral tube, etc.

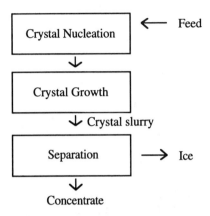

Figure 7.10. Freeze concentration process. (Adapted from Hartel, 1992.)

The condensers are placed within the drying chamber with a suitable radiation shield to minimize radiant heat transfer (Schwartzberg, 1982). Condensers should be arranged so that vapor flow along the surfaces sweeps the noncondensable compounds into the vapor stream removed by the vacuum pump. The latter prevents the buildup of frost in the condenser, which reduces the condensation capacity of the heat exchanger.

The rate of migration of water between the product and the condenser can be expressed in terms of Stefan's equation (Mellor, 1978):

$$G_{\mathrm{m}} = \left(\frac{DMP}{\ell RT} \right) \ln \left[\frac{\left(P - P_{\mathrm{c}}' \right)}{\left(P - P' \right)} \right] \qquad (29)$$

where D is the diffusion coefficient of water vapor, M is the molecular weight of water, ℓ is the distance between the food and the condenser, R is the universal gas constant, T is the absolute temperature, P is the total pressure, P' is the partial pressure of water vapor at the prod-

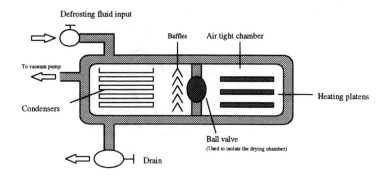

Figure 7.11.   Defrosting arrangement in a freeze dryer. (Adapted from Mellor, 1978.)

uct surface, and P'_c is the partial pressure of water vapor in the vicinity of the condenser. The rate of condensation (G_c) is proportional to the differences between the pressure in the vicinity of the condenser and in the surface:

$$G_c = a'(P'_c - P_c) \tag{30}$$

where a' is a constant and P_c is the vapor pressure of water in the surface of the condenser.

7.4.4 Defrosting

The deposit of frost on the condenser during freeze drying reduces the rate of heat transfer, and thus must be removed or *defrosted.*

The defrosting may be done by passing hot air, hot water, or steam, or using a heating element in the condenser (Mellor, 1978). Figure 7.11 presents an example of a defrosting arrangement in a freeze dryer.

7.5 INDUSTRIAL FREEZE DRYERS

The most commonly used configuration for freeze drying consists of a cabinet with trays connected to a condenser and a vacuum pump. In a batch process, the trays are loaded with the frozen product, the cabinet is closed and evacuated, and the drying temperature is set for the specified duration of the cycle. At the end of the drying cycle, the chamber is filled with an inert gas and then opened. The dried product is then removed and packaged.

Recent advances in industrial practice are focussed on continuous drying methods because they are less labor intensive and less expensive compared to the batch process. There are two types of continuous dryers: tray dryer, where the product is placed on trays and moved along the dryer in a continuous manner; and dynamic or trayless dryers, where the product is moved through the dryer by means of belts, circular plates, vibratory plates, a fluidized bed, and sprays.

Figure 7.12 shows a diagram of a continuous tray dryer, while Figures 7.13 and 7.14 show examples of trayless continuous dryers. Figures 7.15 through 7.17 present some industrial units currently in use for processing freeze-dried products. Large industrial facilities have a freezing room, separated from the drying itself, to speed

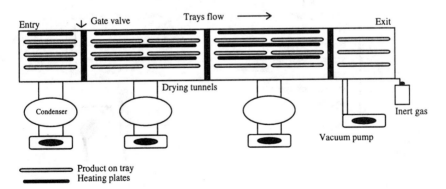

Figure 7.12.   Continuous tray freeze dryer. (Adapted from Karel, 1975.)

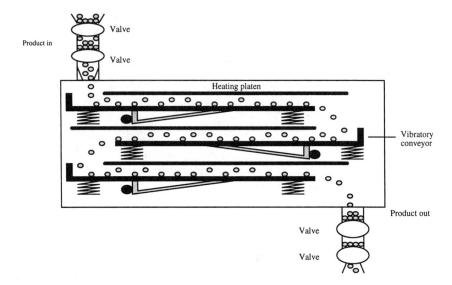

Figure 7.13. **Vibrational freeze dryer (Adapted from Kessler, 1981).**

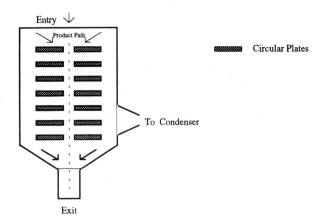

Figure 7.14. **Continuous circular plate freeze dryer. (Adapted from Karel, 1975; Kessler, 1981.)**

Figure 7.15.   Hull® freeze drying system with external condenser. (Courtesy of Hull Corporation, Hatboro, PA.)

Figure 7.16.   Virtis® freeze drying system. (Courtesy of Virtis, Gardiner, NY.)

Figure 7.17.   Virtis® freeze drying system. (Courtesy of Virtis, Gardiner, NY.)

up the initial step of the freeze drying process. The freezing is performed by means of drums or belts (contact cooling) or by trays (cold air circulation—convective cooling). The prefrozen material is then loaded into the dryer and processed. The product will travel in either trays or belts depending on the type of dryer.

7.6 MICROWAVE HEATING IN FREEZE DRYING

The use of microwave heating is widespread today, but the possible corona or glow discharge in a vacuum limited the use of microwave heating in freeze drying (Ma and Peltre, 1973; Dalgleish, 1990). The phenomenon occurs when there is an excess of energy in the drying chamber, which promotes the ionization of the water vapor and results in a spark or glow discharge.

In microwave processing special oscillators known as magnetron, klystrons, and others can be used to generate high frequencies. The magnetron is a cylindrical diode with a ring of resonant cavities that acts as the anode structure. A cavity in the tube becomes resonant or excited in a way that makes it a source for the oscillations of microwave energy. The resonant cavities are vacuum enclosed, and a magnetic field is imposed in the direction parallel to the cylindrical axis. The interaction of electric and magnetic fields results in the development of a space charge. The magnetron oscillates, electrons surrender energy to the microwave field, and a source of power becomes available. Finally, the energy is radiated from an antenna into the food product (Copson, 1962). The power for microwave heating is given by (Keey, 1972; Karel, 1975; Toledo, 1991):

$$p_f = \frac{I_r}{I_c} = E^2 v \epsilon'' \, 55 \times 10^{-14} = \frac{1}{\left(2\pi v C R_s\right)} \tag{31}$$

where v is the frequency, E is the electric field strength, ϵ'' is the loss factor ($\epsilon \tan \delta$), ϵ is the dielectric constant, δ is the complement angle determined from a vector diagram for the total current in a resistance–capacitor circuit, C is capacitance, R is resistance, I_r is the resistance current, and I_c is the capacitor current.

The radiant intensity I at a depth y below the product surface is expressed as:

$$I = I_o \exp\left(\frac{-\lambda_o y}{\left(2\pi \epsilon''\right)}\right) \tag{32}$$

where I_o is the incident intensity and λ_o is the wavelength in free space. The main disadvantages of using microwave heating are the process control, cost, ionization of gases, and glow discharges (Karel, 1975; Dalgleish, 1990).

The design of microwave-heated freeze dryers has to consider a power-to-load matching that leaves no large excess
of energy to support flashover.

The most prevalent settings for microwave heating
energy are 915 MHz and 2450 MHz. Figure 7.18 presents
a general scheme of a microwave freeze dryer. Use of
microwave irradiation offers a method for heating selec-
tively by taking advantage of the difference in dielectric
properties between the ice and the dry portion of the
food. The heating is less dependent on thermal gradient
and thermal conductivity (Copson, 1962).

Arsem and Ma (1990) discussed a mathematical
model intended to describe a combined microwave and
radiant freeze drying process. Most of the works reported
in this area use a pseudo-steady-state approach to
describe the heating in both the frozen and dried regions
(Arsem and Ma, 1990).

The heat balance in the dried and frozen layers can
be expressed as follows (Arsem and Ma, 1990):

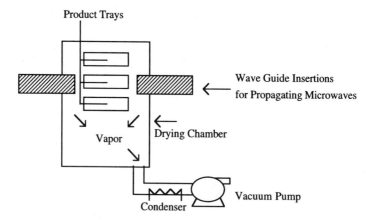

**Figure 7.18   General scheme of a microwave freeze dryer.
(Adapted from Copson, 1962.)**

Dry layer

$$\nabla \cdot (c_{pwv} \, mT_d - K_D \, \nabla T_d) = \omega_d - \rho_d c_{pd} T_d \qquad (33)$$

Frozen layer

$$\nabla \cdot (K_F \, \nabla T_f) = \omega_f - \rho_f c_{pf} T_f \qquad (34)$$

The mass balance on the dry layer can be expresed as:

$$\nabla \cdot (-D_{wv} \, \nabla C_{wv}) = - \sigma C_{wv} \qquad (35)$$

where the diffusion coefficient is expressed in terms of the water vapor pressure as follows:

$$D_{wv} = 78.5 \, / \, (3.4 + P_w) \qquad (36)$$

where K_D and K_F are the thermal conductivities of the dry and frozen layers, c_{pd} and c_{pf} are the specific heat for the dry and frozen layers, C_{wv} is the concentration of water vapor, ω_f is the power generation, ρ_d and ρ_f are the densities of the dry and frozen layers, T_d and T_f are the temperatures of the dry and frozen layers, m is the mass flux, P_w is the partial pressure of water, and σ is the void fraction of the dried layer.

Figure 7.19 shows the temperature profiles within the freeze dryer chamber where the frozen portion has a higher temperature due to microwave irradiation while the heat flows to the drying boundaries by conduction due to the thermal gradient.

Sochanski et al. (1990) reported the freeze dehydration of milk using microwave heating. A comprehensive sublimation model was used to explain moisture content as a function of time and the absorbed microwave energy:

Microwave energy absorbed

$$E_{mw} = S_f \, S\!\!\int\!\!\int q_{mw} \, dz \, dt \qquad (37)$$

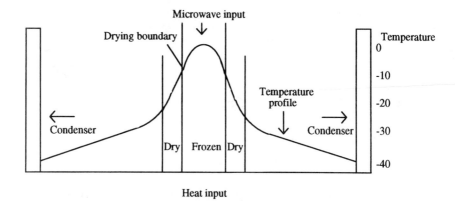

Figure 7.19.   **Temperature distribution through a frozen slab during microwave freeze drying. (Adapted from Charm, 1978.)**

Power generated by microwave

$$q_{mw} = \left(K_f \frac{L}{(2 - x_d)} + K_d x_d \right) E^2 \tag{38}$$

Sublimated water

$$m_{ws} = S_f \int w_{ws}(H, t)\, dt \tag{39}$$

where E is the electric field peak in the sublimation chamber, $x_d \doteq L/2$, S_f is the product layer surface area, S is the product surface/unit volume, L is the product plate thickness, H is the product foam thickness, z is the axial coordinate, t is time, K_f and K_d are the microwave dissipation coefficients for the frozen and dry layers, and w_{ws} is the water flux.

7.7 ATMOSPHERIC FREEZE DRYING

Wolff and Gibert (1989, 1990) discussed the atmospheric freeze drying technique. The process consists of mixing fluidized fine particles of adsorbent and frozen product in a column. Dry and cold gas (air or nitrogen) is used as fluidization media while the column is cooled down. The heat of adsorption provides the required heat for sublimation (Wolff and Gibert, 1989, 1990). Drying is going on automatically when the partial pressure of water and temperature are low enough. A final heating step with hot air is considered to remove the bound water in the product. Figure 7.20 presents an atmospheric freeze drying column as discussed by Wolff and Gibert (1989).

The advantages of atmospheric freeze drying are based on the fluidization properties: good heat transfer coefficient between the immersed product and the adsorbent particles, and the possibility of working in a continuous way. Meanwhile, atmospheric freeze drying has a lower mass transfer rate than vacuum drying.

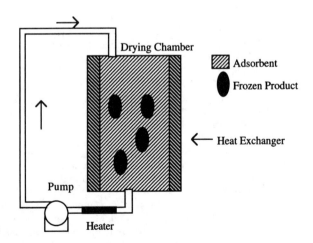

Figure 7.20.   Atmospheric freeze dryer. [Reprinted from Wolff and Gibert (1990) by courtesy of Marcel Dekker Inc.]

7.8 CONCLUDING REMARKS

Freeze drying is a good alternative to preserve foods such as meats, vegetables, and fruits that contain high quantities of proteins or volatiles and that are susceptible to browning reactions. The porous structure resulting from the sublimation of ice allows instant properties in freeze-dried products. Some important parameters to be considered while designing a freeze drying process for a specific product are: the freezing procedure, collapse temperature, secondary drying end point, vacuum, condenser size, and heat source.

The use of microwave irradiation and atmospheric freeze drying are additional alternatives to the conventional operation, and they reduce the operational cost.

7.9 REFERENCES

Arsem, H. B. and Ma, Y. H. 1990. Simulation of a combined microwave and radiant freeze-dryer. *Drying Technol.* 8(5):993–1016.

Bellows, R. J. and King, C. J. 1973. Product collapse during freeze drying of liquid foods. *AIChE Symp. Series 69.* 132:33–41.

Bruin, S. and Luyben, K. Ch. A. M. 1980. Drying of food materials: a review of recent developments. In *Advances in Drying*, Vol. 1, edited by A. S. Mujumdar. Hemisphere Publishing, New York.

Charm, S. E. 1978. Freeze drying. In *Fundamentals of Food Engineering*, Third edition. AVI Publishing, Westport, CT.

Copson, D. A. 1962. Microwave heating. In *Freeze-Drying, Electronic Ovens and Other Applications*, AVI Publishing, Westport, CT.

Dalgleish, J. McN. 1990. *Freeze-Drying for the Food Industries.* Elsevier Applied Science, New York.

Geankoplis, C. J. 1983. *Transport Processes and Unit Operations*, Second edition. Allyn and Bacon, Boston, MA.

Hartel, R. W. 1992. Evaporation and freeze concentration. In *Handbook of Food Engineering*, edited by D. R. Heldman and D. B. Lund. Marcel Dekker, New York.

Heldman, D. R. 1992. Food freezing. In *Handbook of Food Engineering*, edited by D. R. Heldman and D. B. Lund. Marcel Dekker, New York.

Karel, M. 1975. Freeze dehydration of foods. In *Principles of Food Science. Part II: Physical Principles of Food Preservation*, edited by M. Karel, O. W. Fennema, and D. B. Lund. Marcel Dekker, New York.

Karel, M. and Flink, J. M. 1983. Some recent development in food dehydration research. In *Advances in Drying*, Vol. 2, edited by A. S. Mujumdar. Hemisphere Publishing, New York.

Keey, R. B. 1972. *Drying Principles and Practices.* Pergamon Press, New York.

Kessler, H. G. 1981. *Food Engineering and Dairy Technology.* Virlag A. Kessler, Freising, Germany.

Kessler, H. G. 1986. Energy aspects of food preconcentration. In *Concentration and Drying of Foods,* edited by D. MacCarthy. Elsevier Applied Science, New York.

King, C. J. 1970. Freeze-drying of foodstuffs. *CRC Crit. Rev. Food Technol.* 1:379–451.

Longmore, A. P. 1971. Advances in vacuum and freeze drying. *Food Process Ind.* 40:46–49.

Liapis, A. I. and Marchello, J. M. 1984. Advances in the modeling and control of freeze drying. In *Advances in Drying,* Vol. 3, edited by A. S. Mujumdar. Hemisphere Publishing, New York.

Ma, Y. H. and Arsem, H. 1982. Low pressure sublimation in combined radiant and microwave freeze drying. In *Drying'82.* Edited by A. S. Mujumdar. McGraw-Hill, New York.

Ma, Y. H. and Peltre, P. 1973. Mathematical simulation of a freeze drying process using microwave energy. *AIChE Sym. Series 69.* 132:47–54.

Massaldi, H. A. and King, C. J. 1974. Volatiles retention during freeze drying of synthetic emulsions. *J. Food Sci.* 39:438–444.

Mellor, J. D. 1978. *Fundamentals of Freeze-Drying.* Academic Press, New York.

Okos, M. R., Narsimhan, G., Singh, R. K., and Weitnauer, A. C. 1992. Food dehydration. In *Handbook of Food Engineering,* edited by D. R. Heldman and D. B. Lund. Marcel Dekker, New York.

Rey, L. 1975. Some basic facts about freeze drying. In *Freeze Drying and Advanced Food Technology,* edited by S. A. Goldblith, L. Rey, and W. W. Rothmayr. Academic Press, New York.

Sacharow, S. and Griffin, R. C. 1970. *Food Packaging. A guide for the supplier, processor, and distributor.* AVI Publishing, Westport, CT.

Schwartzberg, H. 1982. *Freeze Drying* - Lecture Notes. Food Engineering Department, University of Massachussets, Amherts, MA.

Shukla, K. N. 1990. *Diffusion Processes During Drying of Solids.* World Scientific Publishers, Teaneck, NJ.

Sochanski, J. S., Goyette, J., Bose, T. K., Akyel, C., and Bosisio, R. 1990. Freeze dehydration of foamed milk by microwaves. *Drying Technol.* 8(5): 1017–1037.

Toledo, R. T. 1991. *Fundamentals of Food Process Engineering,* Second edition. Van Nostrand Reinhold, New York.

Van Arsdel, N. B. and Copley, M. J. 1963. *Food Dehydration.* AVI Publishing, Westpoint, CT.

Wolff, E. and Gibert, H. 1989. Atmospheric freeze-drying: experimentation and modelisation. In *Drying'89,* edited by A. S. Mujumdar and M. A. Roques. Hemisphere Publishing, New York.

Wolff, E. and Gibert, H. 1990. Atmospheric freeze drying Part 1. Design, experimental investigation and energy-saving advantages. *Drying Technol.* 8(2):385–404.

CHAPTER 8

OSMOTIC DEHYDRATION

8.0 INTRODUCTION

The concentration of food products by means of product immersion in a hypertonic solution (i.e., sugar, salt, sorbitol, or glycerol) is known as osmotic dehydration (Raoult-Wack et al., 1989; Raoult-Wack, et al., 1991a).

Osmosis consists of the molecular movement of certain components in a solution through a semipermeable membrane to another solution that has a lesser concentration of the particular molecules (Raoult-Wack et al., 1989; Rodríguez-Arce and Vega-Mercado, 1991; Cheryan, 1992; Jayaraman and Das Gupta, 1992). The mechanisms through which osmotic drying of foods is achieved is not simple. Figure 8.1 presents a schematic of the mass transport phenomena during osmotic dehydration.

The water loss during osmotic dehydration can be divided into two periods: (1) an initial period of about 2 h with high rate of water removal and (2) a period that range from 2 to 6 h with a decreasing rate of water removal. The initial rate of water loss is not sensitive to

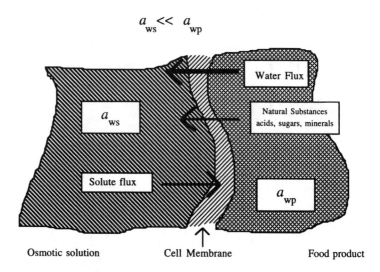

Figure 8.1.   Mass transport during osmotic dehydration. (Adapted from Raoult-Wack et al., 1989.)

the circulation of the osmotic solution. Blanching affects the initial phase of osmotic dehydration, although ultimate water loss is not greatly different from that of the unblanched product. Temperature and concentration of the osmotic solution affect the rate of water loss. Compared to air drying or freeze drying, osmotic dehydration is quicker because the removal of water occurs without a phase change (Farkas and Lazar, 1969; Raoult-Wack et al., 1989; Jayaraman and Das Gupta, 1992).

The process has received considerable attention in past years because of potential industrial applications. In this chapter we discuss the fundamentals of osmotic dehydration and its application in the food industry.

8.1 FUNDAMENTALS

The difference in chemical potential across a semipermeable membrane between product and osmotic solution is the driving force for mass transfer. The water activity is related to the chemical potential, μ_i, as follows:

$$\mu_i = \mu_i^0 + RT \ln a_w \tag{1}$$

where μ_i^0 is the chemical potential of a reference state, R is the universal gas constant, and T is the temperature. The osmotic dehydration proceeds until the water activity of both the solution and the product reaches equilibrium. A comprehensive discussion of water activity is included in Chapter 3.

The initial concentration of the solute (i.e., sugar) in the osmotic solution can be estimated by a simultaneous use of the Ross and Norrish equations in the case of non-electrolytes. The Ross equation can be expressed as follows:

$$a_{w\ equil} = a_{w\ fruit} * a_{w\ sugar} \tag{2}$$

where $a_{w\ fruit}$ is the water activity of the fresh fruit and $a_{w\ sugar}$ is the water activity of the osmotic solution. Both have the same molality and temperature. The amount of sugar required to obtain a given $a_{w\ sugar}$ value can be evaluated from the Norrish equation. The following example illustrates the discussed topic.

EXAMPLE 1

The water activity of fresh pineapples is 0.99 with a moisture content of 80 kg of water/kg of dry solids. The fruit is immersed in a sucrose solution considering a ratio of 4:1 (weight solution/weight fruit) to reduce its water activity to 0.95. Determine the initial water activity of the sucrose solution and the sucrose concentration.

Answer

The initial water activity of the sucrose solution is evaluated from the Ross equation:

$$a_{w\ eq} = a_{w\ pine} * a_{w\ suc}$$

Then, the water activity of the sucrose solution is expressed as:

$$a_{w\ suc} = a_{w\ eq}/a_{w\ suc}$$

By substituting the known values for $a_{w\ eq} = 0.95$ and $a_{w\ pine} = 0.99$ the water activity of the sucrose solution is 0.94. This value is used to calculate the sucrose concentration by means of the Norrish equation:

$$a_{w\ suc} = X_1 \exp(-K(1 - X_1)^2)$$

where K is 6.47 for sucrose. The X_1 value is 0.955 mol of water per total moles of solution or 0.045 mol of sugar per total moles of solution.

8.2 MASS AND HEAT BALANCES

As stated at the beginning of this chapter, water transfer is one of the most important aspects during osmotic dehydration. The main mechanism of mass transfer is diffusion because of the concentration gradient between the osmotic solution and the food.

The rate of water diffusion can be estimated by the modified Fick's Law (Crank, 1956; Conway et al., 1983; Aguerre et al., 1985) as follows:

For a slab geometry

$$\frac{\partial w}{\partial t} = D \frac{\partial^2 w}{\partial x^2} \tag{3}$$

For a cylinder geometry

$$\frac{\partial w}{\partial t} = \frac{1}{r} \frac{\partial}{\partial r}\left(rD \frac{\partial w}{\partial r}\right) \tag{4}$$

where w is the water loss or sugar gain during the osmotic treatment, t is time, D is the diffusion coefficient, x is the thickness of the slab, and r is the diameter of the cylinder. Crank (1956) used Laplace transformation to obtain the following analytical expressions for Eq. (3) and (4), considering an agitated solution of limited volume and the following boundary conditions:

$$\text{at} \quad t = 0 \quad X_t = X_o \quad B = B_0$$

$$t = 0 \quad X_t = X_{eq} \quad B = B_{eq}$$

Water loss or sugar gain on infinite slab

$$\frac{w_{pt}}{w_{p\infty}} = 1 - \Sigma \frac{2\alpha(1+\alpha)}{1+\alpha+\alpha^2 q_n^2} \exp\left(\frac{-Dq_n^2 t}{z^2}\right) \tag{5}$$

Water loss or sugar gain on a cylinder

$$\frac{w_{pt}}{w_{p\infty}} = 1 - \Sigma \frac{4\alpha(1+\alpha)}{4+4\alpha+\alpha^2 q_n^2} \exp\left(\frac{-Dq_n^2 t}{a^2}\right) \tag{6}$$

where w_{pt} is the loss or gain at time t, $w_{p\infty}$ is the loss or gain at equilibrium, α is the fractional uptake, q_n are the zero positive roots of $\tan q_n = -\alpha q_n$ for slab geometry and $\alpha q_n J_0(q_n) + 2J_1(q_n) = 0$ for cylinder geometry, z is half thickness of slice, D is the apparent diffusivity of water, and a is the cross-section of the cylinder. The water loss or sugar gain can be defined as (Conway et al., 1983; Monsalve-González et al., 1993):

$$w_{pt} = \frac{\left(w_o X_o - w_t X_t\right)}{w_o} \tag{7}$$

$$S_g = w_t B_t - w_o B_0 \tag{8}$$

where w_o is the fruit weight at time zero, X_o is the moisture content at time zero, w_t is the fruit weight at time t, X_t

is the moisture content at time t, B_0 and B_t are the °Brix at times zero and t respectively and S_g is used to define the "sugar gain" value. Equations (5) and (6) can be solved using non-linear regression and the apparent diffusivity determined.

The dependence of the diffusion coefficient D on temperature can be expressed as follows:

$$D_{\text{eff}} = D_0 \exp\left(\frac{-E_a}{R_g T}\right) \qquad (9)$$

where E_a is the activation energy, T is the absolute temperature, D_0 is a reference coefficient, and R_g is the universal gas constant. The activation energy can be determined by plotting $\ln D_{\text{eff}}$ versus $1/T$.

Process temperature markedly affects the rate of osmosis (Ponting et al., 1966; Lenart and Flink, 1984a,b). An increase in temperature intensifies the water removal and osmotic substance penetration into the tissue. However, the ratio of removed water to penetrating osmotic substance has a constant value. Also, the temperature

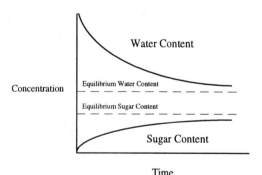

Time

Figure 8.2.   Water and sucrose content during osmotic dehydration. (Adapted from Karel, 1975.)

effect on osmotic kinetics depends on the kind of osmotic substance used (Lenart and Lewicki, 1989).

The water content and sugar concentration as function of time is shown in Figure 8.2. The process duration should be kept as short as possible to achieve a good dehydration. Rapid losses of water are achieved during the first 2 h of an osmotic treatment (Farkas and Lazar, 1969; Lenart and Flink, 1984a; Giangiacomo et al., 1987; Torreggiani et al., 1987). An early interruption of the osmotic process leads to a considerable amount of water removed without great sugar uptake (Karel, 1975).

8.3 OSMOTIC SOLUTIONS

The selection of the solute or solutes for the osmotic solution is based on three main factors: (1) sensory characteristics of the product; (2) the cost of the solute or solutes; and (3) the molecular weight of the solute (Marcotte, 1988). The most common solutes used for osmotic dehydration are sodium chloride, sucrose, lactose, high-fructose corn syrup, and glycerol. Ethanol, alanine, L-lysine, monosodium glutamate, galactose, sodium lactate, polyethylene glycol, casein, serum albumin, soy protein, and glycine are cited in the literature as water activity depressors but not frequently used (Lerici et al., 1985; Chirife and Favetto, 1992). Table 8.1 summarizes the uses and advantages of some osmotic agents.

8.4 HURDLE TECHNOLOGY AND OSMOTIC DEHYDRATION

Most of preservation processes are based on hurdles that affect the conditions where organisms can grow. These changes lead to the inactivation of organisms that are not a health hazard unless the proper growing conditions are restored. Some examples of hurdles are listed in Table 8.2 (Leistner et al., 1981).

Table 8.1   Uses and advantages of osmotic agents.

Name	Uses	Advantages
Sodium chloride	Mainly meats and vegetables Solutions up to 10%	High a_w depression capacity
Sucrose	Mainly fruits	Reduces browning and increases retention of volatiles
Lactose	Mainly fruits	Partial substitution of sucrose
Glycerol	Fruits and vegetables	Improves texture
Combination	Fruit, vegetables, and meats	Adjusted sensory characteristics, combines high a_w depression capability of salts with high water removal capacity of sugar

The factors used to design a process and select the hurdles are based upon the initial amount and type of microorganisms. The main idea in hurdle technology or combined methods is to not use a single factor or hurdle to preserve a food. Instead, several factors are used to increase the energy requirements needed for microbial survival, or decrease the energy sources within the food, as summarized in Figure 8.3. Certain types of foods will be microbiologically stable while they are minimally processed, by lowering pH and water activity, applying a

Table 8.2.   Example of hurdles in food preservation.

Mild heat process
Storage temperature
pH
a_w
Redox potential
Preservatives
Radiation
Competitive flora

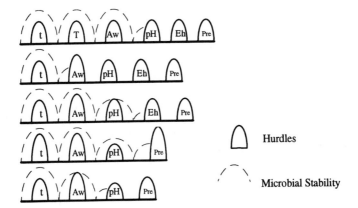

Figure 8.3. Hurdle effect on microbial stability. (Adapted from Leistner et al., 1981.)

mild heat treatment, and adding small amounts of antimicrobial agents.

8.4.1 pH as a Hurdle

Lowering the pH is the most widely used treatment in food preservation (Gould and Jones, 1989). The benefit of lowering pH for the preservation of foods by combined methods is the increase of limiting water activity of bacteria and the decrease of thermal resistance of bacteria, while it potentiates the effect of antimicrobials (Chirife and Favetto, 1992).

Lowering cytoplasm pH and altering metabolism by the acidulants are the main postulated mechanisms of inhibiting microorganisms. Preservatives susceptible to dissociation are thus more effective at a lower pH level of the substrate (Lueck, 1980).

The selection of an acidulant is based on factors such as flavor modification, ability to lower pH, and specific

Effect	Acidulant	Dissociation Constant (pK_a)
Most ↓	Acetic Acid	4.76
	Lactic Acid	3.08
	Citric Acid	3.14
	Malic Acid	3.40
	Tartaric Acid	2.98
Least ↑	Chloridric Acid	Totally Dissociated

Figure 8.4.   Effect of acidulant on *Staphylococcus aureus*. (Adapted from Chirife and Favetto, 1992.)

antimicrobial action (i.e., undissociated molecules and organic acids). The latter is demonstrated in the inhibition of *Staphylococcus aureus*, which depends on the type of acid as presented in Figure 8.4.

8.4.2 Antimicrobials as Hurdles

The antimicrobial activity of weak lipophilic organic acids such as acetate, propionate, sorbate, and benzoate, and the organic acids such as sulfite and nitrite is due, in part, to the increased membrane permeability of their protonated forms (Gould and Jones, 1989). The mode of action of the preservatives ranges from cell membrane interaction to the inhibition of enzymatic activity as summarized in Table 8.3.

Other widely used preservatives are the sulfite compounds which are reported as antimicrobial and antibrowning agents. The SO_2 compounds are reported to have a synergistic effect with sorbic acid (Rose and Pilkington, 1989; Vega-Mercado and Silva-Negrón, 1991; Welti, 1991). Recently, concerns over allergenic reactions in the asthmatic population have prompted the US Food and Drug Administration to limit the use of sulfites in certain food products (i.e., apples).

Table 8.3.   Food preservatives and mode of action.

Preservative	Inhibited Organisms	Mode of Action
Acetic acid CH_3COOH	Bacteria	Interact with the cell membrane. Inhibit the amino acid uptake in membrane vesicles. Decrease the internal pH of cell.
Propionic acid CH_3CH_2COOH	Molds *Bacillus subtilis* Salmonella	Membrane-directed activity. Inhibit the uptake of amino acids in vesicles by the neutralization of protonmotive force by undissociated molecules.
Lactic acid $CH_3CHOHCOOH$	*Mycobacterium tubersulosis* *Bacillus coagulans* Spore forming bacteria	Prevent aflatoxin and sterigmato-cystin formation by fungi. Same mode as acetic and propionic acids.
Sorbic acid $C_6H_8O_2$	Bacteria Mold Yeast	Sorbates affect a number of sites in germinating and developing spores. Inhibit enzymes involved in the tricarboxylic acid cycle (citric acid cycle and Krebs cycle).

From Lueck (1980) and Eklund (1989).

8.4.3 Air Drying, Vacuum Drying and Freeze Drying as Hurdles

Osmotic dehydration is reported as a sound preprocessing step before freeze dehydration, air drying, solar drying, and vacuum dehydration (Ponting et al., 1966; Farkas and Lazar, 1969; Hawkes and Flink, 1978; Islam and Flink, 1982a,b; Moy et al., 1978; Giangiacomo et al., 1987; Torreggiani et al., 1987; Quinteros-Ramos et al., 1993). As a pretreatment to any of the drying techniques previously mentioned, osmosis appears useful as a means of reducing the processing time and energy consumption. It can also improve the sensory characteristics of the final product (Jayaraman and Das Gupta, 1992).

It has been reported that apples with a solids content of 25% to 35%, pretreated with 60% sucrose solution, acquired improved organoleptic properties after freeze drying (Hawkes and Flink, 1978). It is noted that when applying osmotic dehydration prior to freeze drying, the water load is reduced, thus improving the economics of the whole dehydration process. Dixon et al (1976) used vacuum drying on apple slices after an osmotic concentration step, with highly acceptable results. Similar results are reported by Moy et al.(1978) on mangos and papayas. The osmotic dehydration treatment was carried out at 21°C and the vacuum drying (1 to 2 mm Hg) at 60°C. The use of maltodextrin and lactose as osmotic agents resulted in low sweetness levels, which made them useful for products that require less sweetening than fruits (Hawkes and Flink, 1978).

Islam and Flink (1982a) reported a high quality product of potato obtained from air-solar drying. The potato was first dehydrated by osmosis for 4 to 18 h in a 45% sucrose/15% salt solution. It was then solar dried for one day with an air velocity of either 0.47 or 1 m/s. Welti (1991) and Tamayo-Cortes et al (1994) reported the processing of *chicozapote* (mexican fruit) by osmotic dehydration before solar drying.

8.5 APPLICATIONS OF OSMOTIC DEHYDRATION AND HURDLE TECHNOLOGY

A general scheme of an osmotic dehydration system is shown in Figure 8.5. The system consists of a storage tank where the osmotic solution is prepared, followed by a pump and a rotameter to control the flow rate entering the

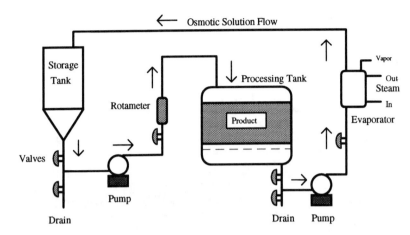

Figure 8.5.   Typical osmotic dehydration system configuration.

processing tank. The product is placed in the processing tank where the osmotic solution is pumped in at a constant rate. Finally, the osmotic solution is removed, concentrated, and returned to the storage tank to reuse. This concept was used by Bajema et al. (1993) in a preliminary design for a pilot plant intended for the processing of apples as shown in Figure 8.6. Raoult-Wack et al. (1989) summarize the application of osmotic dehydration at an industrial level as presented in Figure 8.7. The hurdle technology takes advantage of the osmotic dehydration procedure to incorporate the food additive into the product under treatment. The addition of an antimicrobial agent, antioxidants, and organic acids into the osmotic solution allows not only the removal of water from the product, but also the addition of the food additive to the product.

The use of osmotic dehydration and hurdle technology for processing fruits, vegetables, fish and meat products, and the production of intermediate moisture foods is reviewed in this section. Sometimes osmotic dehydra-

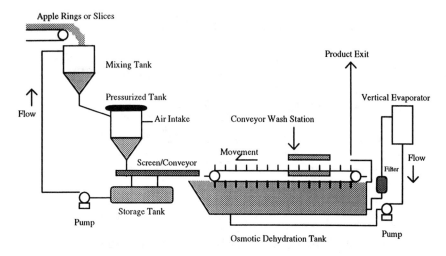

Figure 8.6.   Osmotic dehydration pilot plant. (Adapted from Bajema *et al.*, 1993.)

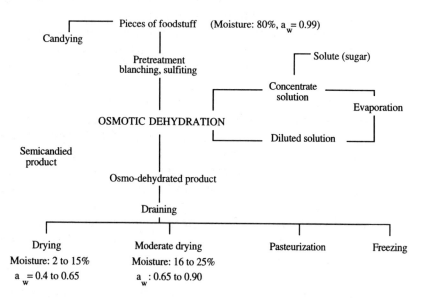

Figure 8.7.   Industrial application of osmotic dehydration. (Adapted from Raoult-Wack et al., 1989.)

tion is used as a preprocessing step prior to a regular drying process. In other instances, the preservation utilizes hurdle technology by reducing the water activity using certain osmotic agents and adding small amounts of antimicrobial agents or changing the pH.

8.5.1 Fruits

Apple products that are pretreated using osmotic dehydration and hurdle technology before freeze or vacuum drying have been reported by several authors such as Ponting et al. (1966), Dixon et al., (1976), Hawkes and Flink (1978), Monsalve-González et al. (1993), Bajema et al. (1993), and Quinteros-Ramos et al. (1993). An improvement in the organoleptic and textural properties and inhibition of nonenzymatic browning were attained using this technique. An additional modification to the regular osmotic dehydration procedure was introduced by Schwartz et al. (1994a). It was the use of concentrated apple juice (67°Brix for 7 h) as an osmotic solution, instead of a sugar solution, in the treatment of apple slices followed by air drying at 65°C until a water activity of 0.51 was reached.

Mango products treated either as slices or puree by hurdle technology are reported by Moy et al. (1978), Ramamurthy et al. (1978), Welti (1991), Rojas et al. (1994), Treviño et al. (1994), and Mata et al. (1994). The preservation of mango puree is attained by adding the preservatives and water activity depressors directly to the mango rather than immersing the fruit in an osmotic solution. The shelf life of the mango mixture ranges from 3 to 6 months at room temperature using this processing technique (Rojas et al., 1994). The treatment of mango slices follows the same procedure previously discussed for apple slices.

Bananas, guava, papaya, *chicozapote*, cherries, passion fruit, tamarind and pineapples have been processed using osmotic dehydration (Ponting et al., 1966; Moy et al., 1978; Alzamora et al., 1989; Vega-Mercado et al., 1991; Vega-Mercado and Silva-Negrón, 1991; Welti, 1991; Garcia et al., 1992; Elguezabal et al., 1994; Schwartz et al., 1994b; Tamayo-Cortes et al., 1994). In each case, the fruit is immersed in an osmotic solution until it reaches a particular water activity level. Pineapple slices stored at room temperature have a shelf life of 30 days to 3 months after being processed by osmotic dehydration. Sulfite, citric acid, and potassium sorbate were used as preservatives whereas sucrose was used as an osmotic agent (Alzamora et al., 1989; Vega-Mercado et al., 1991; Vega-Mercado and Silva-Negron, 1991).

8.5.2 Vegetables

Sodium chloride is used in most vegetable processing because of flavor considerations. Also, a mixture of sucrose and salt is used as an osmotic solution in some applications and the final product has been ranked as acceptable (Lenart and Flink, 1984a; Ibarra et al., 1992). Similar to its use in fruit processing, osmotic dehydration is used as a preprocessing step prior to regular drying, as in the preservation of potatoes using osmotic dehydration prior to air and solar drying (Islam and Flink, 1982a,b).

Other products preserved by hurdle technology are pumpkins and green tomatoes using a sucrose (40% solids) and citric acid (0.2%) solution. The process is carried out for 24 h in a four-step treatment, which leads to products with a shelf life of 12 months (Welti, 1991). Jayaraman and Das Gupta (1992) discussed the treatment of okra, aubergines, tomato, and pepper using sodium chloride (27% solution). The ratio of vegetables to solution plays an important role in the final water activity.

Sliced carrots are processed similarly to potatoes, except for the use of glycerol or propylene glycol as part of the osmotic solution (Kaplow, 1970; Welti, 1991; Jayaraman and Das Gupta, 1992). Process time varies from 2 to 18 h depending on processing factors such as the osmotic solution composition, temperature, and product/solution ratio.

8.5.3 Fish and Meat Products

The use of hurdle technology on mince from pelagic fish results in an extended lag phase, decreased growth phase rate, and lower stationary phase as shown in Figure 8.8 (Aguilera et al., 1993). Sodium chloride is used as a water activity depressant while acetic acid is used to decrease the pH and potassium sorbate is added as a preservative (Del Valle and Nikerson, 1967a,b). In addition to the above factors, a mild heating extends the shelf life of washed minced fish for up to 15 days at 15°C (Aguilera et al., 1993). Fish products may be reformulated by using glycerol, sorbitol, potassium sorbate, salt, corn starch, phosphoric acid, monosodium glutamate, garlic, onion, and guar gum (Aguilera et al., 1993; Morales et al., 1994).

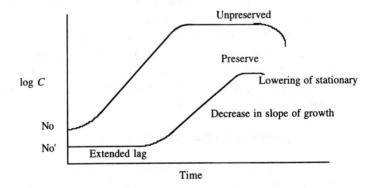

Figure 8.8.   Effect of hurdle technology on microbial growth on fish products (Adapted from Aguilera et al., 1993. Reproduced with permission of the American Institute of Chemical Engineers. Copyright© 1993 AIChE. All rights reserved.)

Table 8.4.   Fish and meat products preserved by combined methods.

Product	A_w	Fat	Low pH	Mild Heat	Preservatives	Reduced Temperature	Competitive Flora	Packaging
Minced fish	X		X	X	X	X		X
Processed meat	X		X		X		X	
Meat spreads	X	X	X		X			X

From Aguilera et al. (1993).

Osmotic pretreatment prior to freeze drying of *Cynoscion maracaibuensis* reduced the drying time by 30% (Romero et al., 1994).

The water activity of meat products ranges from 0.65 to 0.90 (Leistner et al., 1981). The preservation of meat and meat products using hurdle technology or combined methods can be achieved by controlling water activity, pH, E_h (Redox potential); mild heating; and using preservatives (sorbic acid, nitrites, etc.) or competitive flora (*Lactobicillaceae, Streptococcaceae*, molds). Products such as salami are preserved by a combination of curing salts and smoking, low pH, and mild heating (Aguilera et al., 1993). Meat spreads are stabilized by combining salt, phosphoric acid, fat, and potassium sorbate. The use of different humectants is considered while formulating intermediate moisture meat products (Leistner *et al.*, 1981; Aguilera *et al.*, 1993). Table 8.4 summarizes the preservation factor in selected shelf-stable fish and meat products.

8.5.4 Intermediate Moisture Foods (IMF)

Intermediate moisture foods are formulated products that are stable at room temperature without thermal processing and can be eaten without rehydration. The water activity of IMF ranges from 0.65 to 0.85 and the moisture content varies between 15% and 40%. IMF products existed in many countries under different names, formulations, and consumption patterns. The hurdle technology is an extension of the IMF concept where a small depression in water activity is supplemented by other factors such as pH, mild heating, preservatives, and packaging (Aguilera et al., 1993).

8.6 REHYDRATION OF OSMO-DRIED PRODUCTS

The rehydration of osmo-dried products is affected by the osmotic agent and the concentration used in the dehydration step (Moy et al., 1978; Duckworth, 1981). Osmo-dried products tend to gain more water than they originally contained. Vidales et al. (1992) explained the increase in the amount of water adsorbed by osmotically dried strawberries in terms of the complexity of the chemical composition of the fruits, hydrolysis of sucrose, and oversaturation with sugar. These results are similar, to some extent, to those reported by Duckworth (1981) on gelatinized potato starch where the addition of a solute to the gel increased the amount of retained water as shown in Figure 8.9.

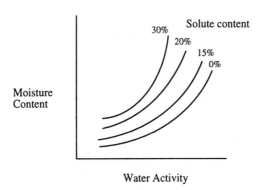

Figure 8.9.   Sorption characteristics of osmotic treated gelatinized potato starch. (Adapted from Duckworth, 1981.)

Rehydrated chicken and carrots, osmotically treated and freeze dried, had no organoleptic differences after 5 minutes in water (Kaplow, 1970). The rehydration properties of celery are improved by osmotic dehydration using sucrose prior to air drying (Neumann, 1972).

8.7 CONCLUDING REMARKS

The increased interest in like-fresh food products makes osmotic dehydration and hurdle technology a good option for food preservation. Further work is necessary to find and develop a humectant that does not alter the flavor of processed products.

The quality of dried products treated by osmotic dehydration followed by another dehydration technique (i.e., freeze drying, vacuum drying, air drying) suggests that future food processing plants will include the possibility of implementing combined dehydration approaches.

8.8 REFERENCES

Aguerre, R. J., Gabitto, J. F., and Chirife, J. 1985. Utilization of Fick's second law for the evaluation of diffusion coefficients in food processes controlled by internal diffusion. *J. Food Technol.* 20:623–629.

Aguilera, J. M., Chirife, J., Parada-Arias, E., and Barbosa-Cánovas, G. V. 1993. CYTED-D AHI: an Ibero-American Project on Intermediate Moisture Foods and Combined Methods Technology. *AIChE Symp. Series 297. Food Dehydration.* 89:72–79.

Alzamora, S. M., Gerschenson, L., Cirrotti, P., and Rojas, A. M. 1989. Shelf-stable pineapple for long term non-refrigerated storage. *Lebensm.-Wiss.u-Technol.* 22:233–236.

Bajema, R. W., Barbosa-Cánovas, G. V., Monsalve-González, A. and Cavalieri, R. P. 1993. Preliminary design for an apple preservation pilot plant utilizing combine methods. In *Boletín Informativo de Divulgación.* 1:46–55. Universidad de las Américas. Puebla, México.

Cheryan, M. 1992. Concentration of liquid foods by reverse osmosis. In *Handbook of Food Engineering*, edited by D. R. Heldman and D. B. Lund. Marcel Dekker, New York.

Chirife, J. and Favetto, G. 1992. *Fundamentals aspects of food preservation by combined methods.* IUFoST short course. Universidad de las Américas. Puebla, México.

Conway, J., Castaigne, F., Picard, G., and Vovan, X. 1983. Mass transfer considerations in the osmotic dehydration of apples. *Can. Inst. Food Sci. Technol. J.* 16(1):25–29.

Crank, J. 1956. *The Mathematics of Diffusion*. Oxford University Press. London, UK.

Del Valle, F. R. and Nikerson, J. T. R. 1967a. Studies on salting and drying fish. I Equilibrium consideration in salting. *J. Food Sci.* 32:173–179.

Del Valle, F. R. and Nikerson, J. T. R. 1967a. Studies on salting and drying fish. II Dynamic aspects of the salting of fish. *J. Food Sci.* 32:218–224.

Dixon, G. M., Jen, J. J., and Paynter, V. A. 1976. tasty apples slices results from combined osmotic-dehydration and vacuum drying process. *Food Prod. Devel.* 10(7):60–64.

Duckworth, R. B. 1981. Solute mobility in relation to water content and water activity. In *Water Activity: Influences on Food Quality*, edited by L. B. Rockland and G. F. Stewart. Academic Press, New York.

Eklund, T. 1989. Organic acids and esters. In *Mechanisms of Action of Food Preservation Procedures*, edited by G. W. Gould. Elsevier Science, New York.

Elguezabal, L., Navarro, P., and Daly, M. 1994. Bulk preservation of three fruits (pineapple, passion fruit and tamarind) by combined methods. In *Boletín Internacional de Divulgación* 2:62–71. Universidad de las Americas, Puebla, México.

Farkas, D. F. and Lazar, M. E. 1969. Osmotic dehydration of apple pieces. Effect of temperature and syrup concentration on rates. *Food Technol.* 23:688–690.

Garcia, A., Vicente, I., Sevillano, E., Castro, D., Acosta, V., Garcia, A., Casals, C. Fernández, C., and Nuñez, M. 1992. Preservation of guava pulp by combined methods. In *Boletín Internacional de Divulgación* 1:31–41. Universidad de las Americas, Puebla, México.

Giangiacomo, R. Torreggiani, D., and Abbo, E. 1987. Osmotic dehydration of fruit. Part 1. Sugar exchange between fruit and extracting syrups. *J. Food Proc. Pres.* 11:183–195.

Gould, G. W. and Jones, M. V. 1989. Combination and synergistic effects. In *Mechanisms of Action of Food Preservation Procedures*, edited by G. W. Gould. Elsevier Science, New York.

Hawkes, J. and Flink, J. M. 1978. Osmotic concentration of fruit slices prior to freeze dehydration. *J. Food. Proc. Pres.* 2:265–284.

Ibarra, E., Bácenas, M. E., López-Malo, A., Argaiz, A., and Welti, J. 1992. Osmotic concentration of potato. In *Boletín Internacional de Divulgación* 1. Universidad de las Americas, Puebla, México.

Islam, M. N. and Flink, J. M. 1982a. Dehydration of potato. I. Air and solar drying at low velocity. *J. Food Technol.* 17:373–385.

Islam, M. N. and Flink, J. M. 1982b. Dehydration of potato. II. Osmotic concentration and its effect on air drying behavior. *J. Food Technol.* 17:373–385.

Jayaraman, K. S. and Das Gupta, D. K. 1992. Dehydration of fruit and vegetables—recent developments in principles and techniques. *Drying Technol.* 10(1):1–50.

Kaplow, M. 1970. Commercial development of intermediate moisture foods. *Food Technol.* 24:889–893.

Karel, M. 1975. Dehydration of foods. In *Principles of Food Science. Part II. Physical Principles of Food Preservation*, edited by M. Karel, O. R. Fennema, and D. B. Lund. Marcel Dekker, New York.

Leistner, L., Rödel, W., and Krispien, K. 1981. Microbiology of meat and meat products in high and intermediate moisture ranges. In *Water Activity: Influences on Food Quality*, edited by L. B. Rockland and G. F. Stewart. Academic Press, New York.

Lenart, A. and Flink, J. M. 1984a. Osmotic concentration of potato. I. Criteria for the end-point of the osmosis process. *J. Food Technol.* 19:45–63.

Lenart, A. and Flink, J. M. 1984b. Osmotic concentration of potato. II. Spatial distribution of the osmotic effect. *J. Food Technol.* 19:65–89.

Lenart, A. and Lewicki, P. R. 1989. Osmotic dehydration of apples at high temperature. In *Drying '89*, edited by A.S. Mujumdar and M. Roques. Hemisphere Publishing, New York.

Lerici, C. R., Pinnavaia, G., Dalla Rosa, M., and Bartolucci, L. 1985. Osmotic dehydration of fruit: influence of osmotic agents on drying behavior and product quality. *J. Food Sci.* 50:1217–1219, 1226.

Lueck, E. 1980. *Antimicrobial Food Additives*. Translated by G. F. Edwards. Springe-Verlag, New York.

Marcotte, M. 1988. Mass transport phenomena in osmotic processes; experimental measurements and theoretical considerations. Master's Thesis, The University of Alberta, Department of Food Science, Edmonton, Alberta.

Mata, M. M., Tovar, G. B., and Castillo, C. M. J. 1994. Elaboration of a mango fruit pasta. In *Boletín Internacional de Divulgación* 2:52–55. Universidad de las Americas, Puebla, México.

Monsalve-González, A., Barbosa-Cánovas, G. V., and Cavalieri, R. P. 1993. Mass transfer and textural changes during processing of apples by combined methods. *J. Food Sci.* 58(5):1118–1124.

Morales, L. J., Santillán, M., and Orozco, G. 1994. Elaboration at pilot plant level on an intermediate moisture fish hamburger. Presented at the *International Symposium on the Properties of Water. Practicum II*. Universidad de las Américas, Puebla, México.

Moy, J. H., Lau, N. B. H., and Dollar, A. M. 1978. Effects of sucrose and acids on osmovac-dehydration of tropical fruits. *J. Food Proc. Pres.* 2:131–135.

Neumann, H. J. 1972. Dehydrated celery: effect of pre-drying treatment and rehydration procedures on reconstitution. *J. Food Sci.* 37:437–441.

Ponting, J. D., Watters, G. G., Forrey, R. R., Jackson, R., and Stanley, W. L. 1966. Osmotic dehydration of fruits. *Food Technol.* 20:125–128.

Quinteros-Ramos, A. De la Vega, C., Hernández, E., and Anzaldúa-Morales, A. 1993. Effect of the conditions of osmotic treatment on the quality of dried apple discs. *AIChE Symp. Series 297.* 89:108–113.

Ramamurthy, M. S., Bongirwar, D. R. and Bandyopadhyay, C. 1978. Osmotic dehydration of fruits: possible alternative to freeze-drying. *Indian Food Packer* 32(1):108–112.

Raoult-Wack, A. L., Lafont, F., Ríos, G., and Guilbert, S. 1989. Osmotic dehydration. Study of mass transfer in terms of engineering properties. In *Drying '89*, edited by A.S. Mujumdar and M. Roques. Hemisphere Publishing, New York.

Raoult-Wack, A. L., Guilbert, S., Le Maguer, M., and Ríos, G. 1991a. Simultaneous water and solute transport in shrinking media—Part 1. Application to dewatering and impregnation soaking process analysis (osmotic dehydration). *Drying Technol.* 9(3):589–612.

Raoult-Wack, A. L., Petitdemange, F., Giroux, F., Guilbert, S. Ríos, G., and Lebert, A. 1991b. Simultaneous water and solute transport in shrinking media—Part 2. A compartmental model for dewatering and impregnation soaking process. *Drying Technol.* 9(3):613–630.

Raoult-Wack, A. L., Botz, O., Guilbert, S., and Ríos, G. 1991c. Simultaneous water and solute transport in shrinking media—Part 3. A tentative analysis of spatial distribution of impregnating solute in model gel. *Drying Technol.* 9(3):631–641.

Rodríguez-Arce, A. L. and Vega-Mercado, H. 1991. Osmotic drying kinetics of pineapple and papaya. *J. Agric. Univ. Puerto Rico.* 75(4):371–382.

Rojas, R., Quintero, J. S., Rojas, J. C., and Coronado, B. D. 1994. Unsweetened mango puree preserved by combined factors in large containers. In *Boletín Internacional de Divulgación* 2:45–51. Universidad de las Americas, Puebla, México.

Romero, D., González, L., and Mogollón, L. 1994. Combined osmotic and freeze-drying of fish. Presented at the *International Symposium on the Properties of Water. Practicum II.* Universidad de las Américas, Puebla, México.

Rose, A. H. and Pilkington, B. J. 1989. Sulfite. In *Mechanisms of Action of Food Preservation Procedures*, edited by G. W. Gould. Elsevier Science, New York.

Schwartz, M., Silva, C., and Vergara, P. 1994a. Osmotic dehydration of Granny Smith apples using apple juice and hot air. Presented at the *International Symposium on the Properties of Water. Practicum II.* Universidad de las Américas, Puebla, México.

Schwartz, M., Silva, C., and Villanueva, L. 1994b. Drying bananas by osmosis and hot air. Presented at the *International Symposium on the Properties of Water. Practicum II.* Universidad de las Américas, Puebla, México.

Tamayo-Cortes, J., Osalde-Balam, M., Rivas, Ruiz, J., and Saurí-Duch, E. 1994. Osmotic dehydration of chicozapote (*Achras sapota*) pieces using 3 processes. Sensory evaluation and characteristics of the obtained products. In *Boletín Internacional de Divulgación* 2:18–23. Universidad de las Americas, Puebla, México.

Torreggiani, D., Forni, E., and Rizzoto, A. 1987. Osmotic dehydration of fruit. Part 2. Influence of the osmotic time on the stability of processed cherries. *J. Food Technol.* 12:27–44.

Treviño, T. S., Tovar, G. B., Gutiérrez, M. P., and Rodríguez, D. A. 1994. Preservation of mango puree using the combined factors method. In *Boletín*

Internacional de Divulgación 2:35–44. Universidad de las Americas, Puebla, México.

Vega-Mercado, H. and Silva-Negrón, L. 1991. Research note: Efectividad de preservativos durante el almacenamiento en productos de piña de humedad intermedia. *J. Agric. Univ. Puerto Rico.* 75(4):411–415.

Vega-Mercado, H., Beauchamp de Caloni, I, Díaz, N., and Cruz, J. R. 1991. Efecto de la escaldadura en la vida útil y aspectos químicos de los productos de piña de humedad intermedia. *J. Agric. Univ. Puerto Rico.* 75(1):25–36.

Vidales, S. L., Malec, L.F., Gerschenson, L.M., and Alzamora, S. M. 1992. Water sorption characteristics of sugar osmotically dried strawberries. In *Boletín Internacional de Divulgación* 1:73–78. Universidad de las Americas, Puebla, México.

Welti, J. 1991. *Programa de ciencia y tecnología para el desarrollo V Centenario CYTED-D. Desarrollo de alimentos de humedad intermedia importantes para Iberoamérica. Subproyecto frutas y hortalizas.* Univesidad de las Américas, Puebla, México.

CHAPTER 9

OTHER METHODS OF DEHYDRATION OF FOODS AND PACKAGING ASPECTS

9.0 INTRODUCTION

There are other drying techniques in addition to the ones already discussed in Chapters 5, 6, 7, and 8 based on similar concepts and used in the food industry. This chapter covers drying techniques such as sun drying, vacuum drying, drum drying, microwave drying, extrusion cooking, fluidized bed drying, and pneumatic drying. In addition, this chapter includes some important aspects related to the packaging of dried foods: mechanical damage, permeability, temperature effect, light transmission and, compatibility of packaging materials.

9.1 SUN DRYING

The path of solar energy through the atmosphere is summarized in Figure 9.1 as discussed by Norton (1991). Air molecules, water droplets, dust, and clouds cover scatter and absorb part of the solar radiation as it enters the atmosphere, which results in a reduction of direct solar radiation reaching the surface of the earth. Climatic zones

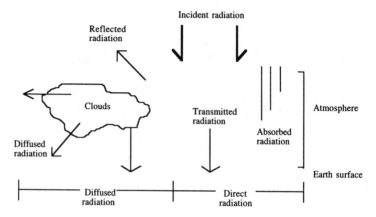

Figure 9.1.   Passage of solar light through the atmosphere. (Adapted from Norton, 1991.)

identified around the world are classified as tropical, semiarid, arid or desert, cold, summer dry, and tundra based on a characteristic insolation level, which is important when designing a solar dryer.

In a solar collection system, the critical insolation is defined at a given ambient temperature as the condition where the heat gain is equal to the heat loss:

$$q_{net} = 0 = F_r A I (\tau\alpha)_e - UA(T_c - T_a) \tag{1}$$

or

$$I_c = \frac{U(T_c - T_a)}{F_r(\tau\alpha)_e} \tag{2}$$

where I is the radiation, T_c is the critical isolation temperature where the heat loss is equal to the heat gained, T_a is the ambient temperature, $(\tau\alpha)_e$ is the effective transmittance–absorbance, A is the surface area, U is the overall heat transfer coefficient and F_r is the radiation coefficient.

The critical heat associated with the insolation can be expressed as:

$$q_c = A F_r (\tau\alpha)_e \, I_\beta \underset{N}{\sum} \left(\frac{I_\beta - I_c}{I_\beta} \right) \tag{3}$$

where I_β is the average insolation of N days, I_β is the actual insolation per day, $(\tau\alpha)_e$ is the average effective transmitance–absorption, and β is the tilt of the system. The *utilisability factor*, ϕ, for solar radiation is defined as follows (Norton, 1991):

Hourly

$$\phi_i = \frac{1}{N} \underset{N}{\sum} \left(\frac{I_\beta}{I_\beta} - \frac{I_c}{I_\beta} \right) \tag{4}$$

Daily

$$\phi(I_c) = \frac{\underset{N}{\sum} \underset{n}{\sum} (I_\beta - I_c)}{\underset{N}{\sum} \underset{n}{\sum} I_\beta} \tag{5}$$

The solar radiation utilisability depends on insolation for a given location, month, and β. Figure 9.2 presents the total daily insolation in different countries around the world.

The heat required for drying on a solar collector for a certain period of time can be estimated as follows (Bansal and Garg, 1987):

$$\eta_1 \eta_2 A I = \frac{Y (\omega_1 - \omega_2) \lambda}{t}$$

$$\eta_1 = 0.33 \eta_d$$

$$\eta_2 = 0.33 \eta_s \tag{6}$$

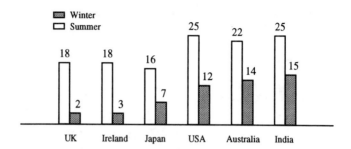

Figure 9.2.   Total daily insolation of selected countries. (Adapted from Norton, 1991.)

where η_d is the drying efficiency, η_s is the efficiency of collecting solar energy, A is the dryer surface area, I is the solar insolation, Y is the yield of crop/hectare, ω_1 is 20% of the initial moisture content, ω_2 is 10% of the final moisture content, λ is the latent heat of vaporization, and t is the harvesting time.

The practice of drying food crops by spreading the food in the open sun in thin layers is termed *open sun drying* or *natural sun drying* (Garg, 1987). The technique is used in the processing of grapes, figs, prunes, coffee beans, cocoa, apricots, plums, pimientos, peppers, rice, and soy beans, among others (Bansal and Garg, 1987; Garg, 1987; Tsamparlis, 1990).

The main limitations of open solar drying are (Bansal and Garg, 1987; Garg, 1987):

- lack of control over the drying process which may result in overdrying, the loss of germination in grains, and nutritional changes.
- lack of uniformity
- contamination by fungi, bacteria, rodents, birds or insects.

Modern solar dryers have been developed on the basis of the concepts utilized with conventional dryers.

Solar radiation as an energy source is also added, which provides an improved operation (Bansal and Garg, 1987; Garg, 1987; Tsamparlis, 1990).

Two main classifications used for solar dryers are: (1) natural convection which does not require any mechanical or electrical power, and (2) forced convection which requires the use of a fan or blower to pump the air (Bansal and Garg, 1987). Both types of dryers are used in processing fruits, vegetables, grains, and cereals, as shown in Table 9.1.

The most common dryers found in the food industry are the direct and ambient types from the natural convection classification, as well as the indirect forced convection type.

9.1.1 Natural Convection or Direct Type

This type of dryer does not use any fan or electrical blower. The drying rate is slow and there is not much

Table 9.1.   **Types of dryers and associated uses.**

Convection	Natural			Forced		
Drying Mode		Direct	Indirect		Direct	Indirect
Heating	Ambient	Preheat Solar	Preheat	Ambient	Preheat	Preheat Solar
Crops	Fruits Vegetables Coffee	Paddy Corn	Paddy Vegetables	Corn Grains	Fruits Cereals Grains Vegetables	Corn Grains
Type	Racks Chambers	Chambers	Bins	Bins	Chambers	Bins

Ambient, Atmospheric air not pretreated; preheat, atmospheric air heated by electricity or fuel; preheat solar, atmospheric air heated by solar energy; direct, product exposed to sunlight; indirect, product exposed to hot air and the whole unit exposed to sun.
Adapted from Bansal and Garg, 1987.

control of temperature and humidity. Only a small amount of product can be processed and some products change in color and flavor (Garg, 1987). The food is heated by direct insolation or evaporation driven by the temperature gradient inside the dryer enclosure.

The rack dryer consists of a certain number of shelves under a metal roof, used to protect the product against rain or excessive sun, as shown in Figure 9.3a. The cabinet solar dryer is based upon the concepts discussed earlier in this book regarding tray or cabinet dryers, as shown in Figure 9.3b. Finally, the greenhouse type, which consists of a structure similar to a greenhouse with shutters that allow ventilation, is shown in Figure 9.3c.

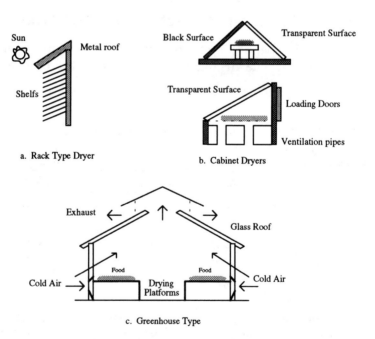

Figure 9.3.   Natural convection solar dryers. (Reprinted from Garg (1987) by permission of Kluwer Academic Publishers.)

9.1.2 Indirect Natural Convection Dryer

This type of solar dryer is superior and more expensive than the direct type since the temperature, humidity, and drying rate can be controlled. The dryer configuration may be like a tray dryer, belt dryer, tunnel dryer, or bin dryer. Figure 9.4 presents some examples of this type of dryer as discussed by Garg (1987) and Bansal and Garg (1987).

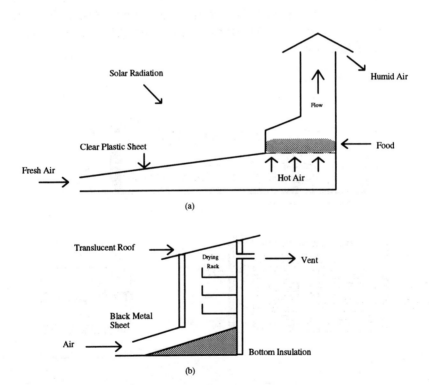

Figure 9.4.   Indirect natural convection dryers: (a) chimney paddy (Reprinted from Garg (1987) by permission of Kluwer Academic Publishers); (b) fruit and vegetable dryer (Adapted from Bansal and Garg, 1987).

9.1.3 Wind-Ventilated or Forced Convection Solar Dryer

This type of dryer uses some kind of fan or blower for the circulation of air. Large amounts of agricultural products can be dried quickly. The dryers can be designated into two classifications: (1) direct mode circulation and (2) indirect mode forced circulation. Figures 9.5 and 9.6 present some examples of forced convection dryers as discussed by Goswami et al (1991), Imre et al. (1990), Rosselló et al. (1990), Stehli and Escher (1990), Norton (1991), Garg (1987), Bansal and Garg (1987), and Tsamparlis (1990).

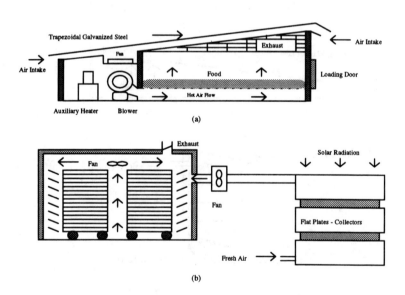

Figure 9.5. Forced convection dryers: (a) layer type dryer (reprinted from Imre et al. (1990) by courtesy of Marcel Dekker Inc.); (b) cabinet type dryer (reprinted from Tsamparlis (1990) by courtesy of Marcel Dekker Inc.)

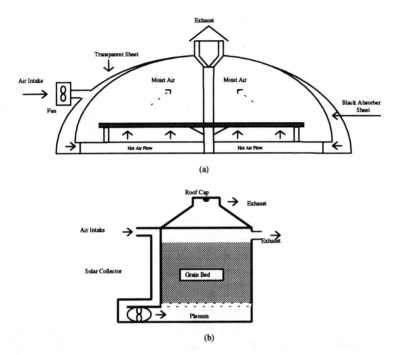

Figure 9.6.   Forced convection dryers: (a) geodesic dome fruit dryer (reprinted from Goswami et al. (1991) by courtesy of Marcel Dekker Inc.); (b) drying bin (reprinted from Garg (1987) by permission of Kluwer Academic Publishers).

Most of the designs discussed in this chapter are used for fruits and vegetables. High-quality products are reported for use of those dryers that provide, in some instances, temperature and humidity control for the incoming air.

9.2 VACUUM DRYERS

Vacuum dryers are used for processing heat sensitive products (Spotts and Waltrich, 1977; Geankoplis, 1983).

Foods in the form of slurries, granules, and sheets are handled in vacuum dryers in batch or continuous operations. They are known as indirect dryers because they use little or no air. Heat transfer occurs through conduction and radiation. Compared to direct dryers, in which the product is in direct contact with the drying medium, the vacuum dryer has a lower maximum drying temperature and lower maximum throughputs. The constant rate drying time can be determined as follows (Charm, 1978):

$$\frac{dX}{dt} = h_t A \left(T_m - T_s \right) \tag{7}$$

$$h_t = \left(h_c + h_r \right) \left[1 + \frac{K A_u}{L \left(H_c + h_r \right)} \right] \tag{8}$$

where X is the moisture content, t is the time, h_r and h_c are the radiation and conduction heat transfer coefficients, A_u is the ratio of exposed surface to nonexposed surface, L is the depth of material, K is the conductivity of material, T_m is the heater temperature, T_s is the vaporization temperature for free water under the vacuum condi-

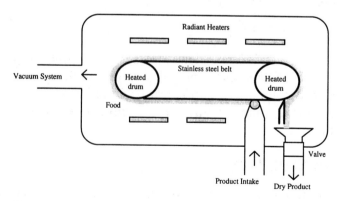

Figure 9.7.   Vacuum band dryer. (Adapted from Brennan *et al.*, 1990.)

tion. The equations derived for freeze drying may be used to describe vacuum drying, considering that in this case the water is not frozen. Figure 9.7 presents an example of a vacuum band dryer as discussed by Brennan et al. (1990). Most vacuum applications use levels of absolute pressure of 50 mm Hg (7 kPa), which lowers the boiling point of water to 39°C.

9.3 DRUM DRYERS

This type of dryer consists of hollow metal cylinders that rotate on horizontal axes and are heated internally by steam, hot water, or other heating medium. Drum dryers are suitable for slurries or pastes in fine suspension, and also for solutions (Geankoplis, 1983). Drum dryers are classified into three types: single drum, double drum, and twin drums. Drums have to be carefully constructed into a perfectly cylindrical shape. Potato flakes are produced using drum dryers. There are various type of drum dryers as shown in Figure 9.8.

The overall drying rate of the food film over the drum can be expressed as follows (Brennan et al., 1990):

$$\frac{dX}{dt} = \frac{F_s\left(X_o - X_f\right)}{t} = \frac{K_c A\left(T_w - T_e\right)}{\lambda} \tag{9}$$

or using the mean temperature difference, ΔT_m, between the roller surface and the product (Heldman and Singh, 1981):

$$\frac{dX}{dt} = \frac{K_c A \Delta T_m}{\lambda} \tag{10}$$

where X_o is the initial moisture content, X_f is the final moisture content, F_s is the mass of solids in the feed, t is time, K_c is the overall heat transfer coefficient, A is the drying surface area, T_w is the temperature of the heated

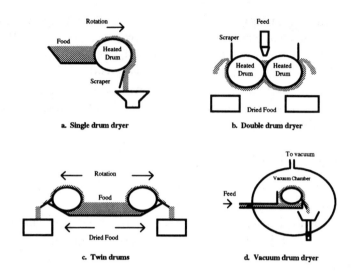

Figure 9.8.   Drum dryers. (Adapted from Karel, 1975 and Brennan et al., 1990).

surface, T_e is the temperature of the evaporating surface, and λ is the latent heat at T_e.

The most important aspect considered when using drum dryers is the uniform thickness of the film applied to the drum surface. This is in addition to the speed of rotation and heating temperatures. All of these affect the drying rate of a drum dryer. The handling of heat-sensitive materials while using drum dryers requires the drums to be enclosed in a vacuum-tight chamber. The main advantages of drum drying are the high drying rates and economic use of heat. The main limitations of drum drying are that the applied food has to be in a liquid or slurry form and it must withstand a relatively high temperature for short periods of time.

Examples of drum-dried products are milk, soup mixes, ingredients for baby foods, potato slurries, and instant cereals.

9.4 MICROWAVE DRYING

Microwaves are high-frequency waves, where the energy strikes an object and is reflected, absorbed, or transmitted through the object. Factors such as dielectric coefficient, shape, and moisture have to be considered while drying a food by means of microwave heating. The advantages of microwave heating over convection or conduction heating are:

- Energy absorption only by the product to be heated.
- Negligible losses on heat transfer media such as air or oven walls.
- Heat can be turned off/on instantaneously.
- Depth penetration of the heat source, which gives more effective and uniform heating.

Specific uses of microwave heating in the food industry are: drying of potato chips, blanching of vegetables, speed thawing of frozen fish, precooking of chicken, and suppression and elimination of molds in dried fruits and dairy products. Sochanski et al. (1990) reported the use of microwave heating in the drying of foamed milk.

Arsem and Ma (1990) expressed the energy balance on a dried layer considering conduction, convection, and volumetric heat generation by microwave irradiation as follows:

$$\nabla(C_{pw}mT_d - K_d \nabla T_d) = \varpi_d \rho_d C_{pd} T_d \tag{11}$$

and the mass balance:

$$\nabla(-D_w \nabla C_w) = -\sigma C_w \tag{12}$$

$$D_w = 78.5(3.4 + P) \quad \text{(for a vacuum chamber)} \tag{13}$$

where C_{pw} is the specific heat of water, C_{pd} is the specific heat of the dried layer, m is the mass flux, T_d is the temperature of the dry layer, H_d is the dissipation coefficient,

ϖ_d is the power generation, ρ_d is the density of the dry layer, D_w is the diffusion coefficient, C_w is the water concentration, P is the pressure, and σ is a constant. Brennan et al. (1990) mentioned partial dehydration of cereals, peas, and beans in bulk using microwave heating. In Chapter 7 we have discussed the application of microwaves on freeze-drying.

9.5 EXTRUSION COOKING

Extrusion is defined as a process whereby food is forced through a die, producing the mixing and forming of the food. The process is well established in the manufacture of pasta and sausages (Harper, 1989; Brennan et al., 1990). Food undergoing extrusion under controlled heating conditions is called extrusion cooking, which is a high temperature–short time (HTST) heating process. Single-screw cooking extruders were developed in the 1940s to process cereal flours or grits and twin-screw extruders were introduced in the 1960s. During the process, the mechanical energy required to turn the screw is converted to heat, raising the temperature of the mixture and resulting in a plasticized feed. Then, the mixture is forced through a die which causes puffing while water vapor is released from the food. Semirigid or rigid products with the cross-sectional profile of the die are obtained when the product passes through the die, expands rapidly, and cools down. The loss in moisture is due to the sudden pressure drop. Food undergoing extrusion is subject to reactions and physical phenomena such as hydration, shearing, homogenization, starch gelation, protein denaturation, melting, plastification, inactivation of microorganisms, forming, shaping, expansion, and drying. All of these processes occur both simultaneously and sequentially. The application of this process is widespread, covering the production of breakfast foods, snack and pet foods, precooked infant foods, flatbreads, crackers, and biscuits. Filled products such as snacks containing more than one component may be produced by coextrusion.

9.5.1 Single-Screw Extruders

A preconditioner section is used to increase the residence time, reduce mechanical power consumption, and increase capacity in extruders. The raw food ingredients are uniformly moistened and heated before they enter the extruder. The extruder consists of three sections: feeding, transition, and metering as presented in Figure 9.9. The extrusion screw sequentially conveys and heats food ingredients and transforms them into a plasticized mass while rotating in a tightly fitting barrel. The characterization of the extrusion operation is expressed in terms of net mechanical energy input divided by mass flow rate. The operation of a single-screw extruder depends on the following variables: pressure drop across the die, slip at the barrel wall, temperature, feed rate, screw speed, melt characteristics, and viscosity of feed (Harper, 1989). The shear rate is use to characterize the single screw extruder:

$$\dot{\gamma} = \frac{dV}{dh} \approx \frac{D\omega}{2H} \tag{14}$$

where dV/dh is the change in velocity with respect to height, $D\omega/2$ is the maximum tip velocity of the screw, and H is the flight height. The strain, γ, is the relative displacement of a fluid element under shear at any time t:

$$\gamma = \dot{\gamma} t \tag{15}$$

Also, the energy input is used to differentiate extruders:

$$E = \frac{W + q + \dot{m}_s \lambda}{\dot{m}} \tag{16}$$

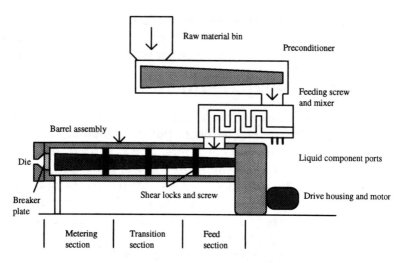

Figure 9.9.   Schematic of an extrusion cooker.

where W is the viscous dissipation of the net mechanical energy used to turn the screw, q is the heat transfer through the jackets in the barrel, l is the latent heat from the steam, ṁ *is the product flow rate, and* s is the steam flow rate applied directly to the ingredients.

9.5.2 Twin-Screw Extruders

A second type of extruders is known as twin-screw. The relative direction of the rotation of the screws, counter or corotating, and the degree of screw intermeshing are key points of differentiation (Harper, 1989). Figure 9.10 summarizes various screw configurations used in twin-screw extruders (Harper, 1989) and Figure 9.11 presents an example of an industrial extruder from Werner & Pfleiderer®.

The counterrotating twin-screw extruders consist of two screws which may or may not be fully intermeshing and rotate in opposite directions. The screws require low

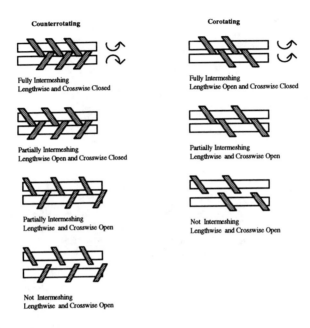

Counterrotating

Fully Intermeshing
Lengthwise and Crosswise Closed

Partially Intermeshing
Lengthwise Open and Crosswise Closed

Partially Intermeshing
Lengthwise and Crosswise Open

Not Intermeshing
Lengthwise and Crosswise Open

Corotating

Fully Intermeshing
Lengthwise Open and Crosswise Closed

Partially Intermeshing
Lengthwise and Crosswise Open

Not Intermeshing
Lengthwise and Crosswise Open

Figure 9.10.   Screw configurations used in twin-screw extruders. (Adapted from Harper, 1989.)

speed to reduce the separating forces and wear caused by the calendering effect at the nip between the screws.

The corotating extruders have become popular in food processing because of their high capacity and enhanced mixing capability. The mixing capability of this type of extruder enhances heat transfer to viscous food materials.

A twin-screw extruder offers further advantages because the screw's greater conveying angle and self wiping feature make it possible to handle a wider variety of ingredients. Also, this type of extruder has fewer interaction of process variables than a single extruder, which makes it easier to operate and control.

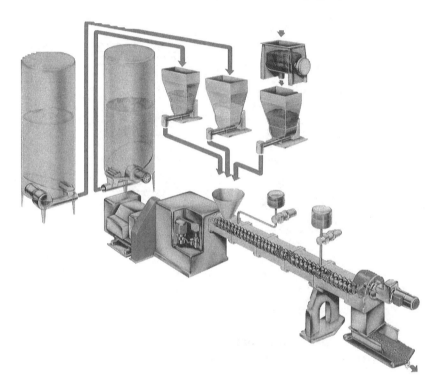

Figure 9.11.   Twin-screw extruder from Werner & Pfleiderer®. (Courtesy of Werner & Pfleiderer®, Ramsey, NJ.)

The use of extruders in the food industry includes cooking and forming pet food (dry and semimoist) using single-screw units, and producing ready-to-eat cereals, snacks, confectionery products, texturized vegetable protein, and macaroni using twin-screw units.

9.6 FLUIDIZED-BED DRYERS

Food particles are fluidized when the pressure drop across a particles bed balances the weight of the particles and the bed expands. The expansion results in the suspension of the particles in the air. The system behaves as a fluid when the Froude number is below unity (Karel, 1975):

$$\text{Froude} = \frac{U^2}{2gr} \tag{17}$$

where U is the air velocity, g is the gravitational constant, and r is the particle radius. Usually the air velocity is in the range of 0.05 to 0.75 m/s. The use of vibratory conveyors assists in maintaining particles suspended for products that are not easily fluidized (Heldman and Singh, 1981). The design of industrial fluidized beds is based on pilot plant testing and on-site operational experience (Masters, 1993). The advantages of fluidized bed technology include both batch and continuous modes, large- and small-scale operations, equipment items with few mechanical moving parts, rapid heat and mass transfer rates between the product and the drying medium, and rapid mixing of solids which leads to nearly isothermal conditions throughout the fluidized layer. They are used in the dairy, food, and pharmaceutical industries for drying, cooling, coating, and agglomeration.

The fundamental aspect of fluidization and fluidized bed operations involves mixing, entrainment, segregation, and heat transfer. Materials suitable for fluidized bed operations can be somewhere between 20 µm and 10 mm to avoid excessive channeling and slugging with a narrow particle size and regular shape. The particles cannot be sticky at the processing temperature (Masters, 1993).

9.6.1 Batch Fluidized Bed

This type of dryer is used extensively where the capacity is small and the quality assurance procedures are a concern. In a batch fluidized bed, the material to be dried is placed in a suitable container with a distribution plate (wire mesh supporting screen) and is subjected to a stream of heated air at a selected temperature until the desired moisture level is reached. The typical configuration of a batch type fluid bed dryer is shown in Figure 9.12 and an example is shown in Figure 9.13. The temperature may be constant or reduced during the drying period (Williams-Gardner, 1971).

9.6.2 Continuous Fluid Bed Dryers

Fluidized solids show similar flow characteristics of those to a liquid. Based on this similarity, the design of a fluidized bed dryer can include an overflow discharge pipe and a supporting grid base through which the fluidizing air is introduced and the fluidized solids can be continuously withdrawn at a controlled rate, as shown in Figures 9.14 and 9.15.

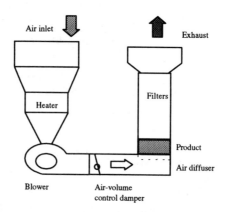

Figure 9.12.   Batch fluidized bed dryer. (Adapted from Williams-Gardner, 1971).

Figure 9.13. Glatt® batch fluidized-bed unit. (Courtesy of Glatt Air Techniques Inc. Ramsey, NJ)

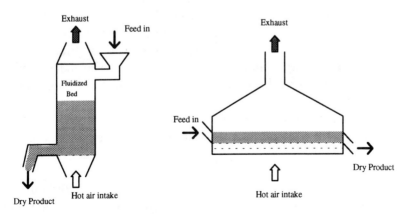

Figure 9.14. Continuous fluidized bed dryers.

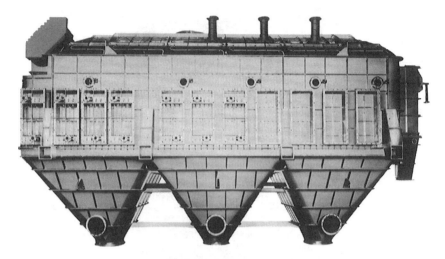

Figure 9.15.   Carrier® continuous fluidized-bed dryer. (Courtesy of Carrier Vibrating Equipment, Inc. Louisville, KY)

9.7 PNEUMATIC DRYERS

This type of dryer is used for drying granular, flaky, or powdery products. Heated air carries the product through the drying zone and into the separation unit. Pneumatic dryers are particularly useful when most of the water is removed in the constant rate drying period (Heldman and Singh, 1981). A pneumatic drying facility consists of a hot air source, material feeding device, the main drying chamber, a cyclone for material–air separation, and a fan, as shown in Figure 9.16.

Assuming that all moisture is on the surface and the particles are spheres, the heat transfer is expressed as:

$$q = U_s A \Delta T_m \qquad (18)$$

where U_s is the surface heat transfer rate, A is the surface area and ΔT_m is the log mean temperature difference. The total heat requirement is expressed as:

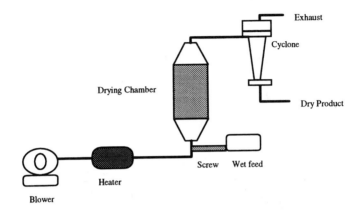

Figure 9.16.   Schematic of a pneumatic drying facility.

$$Q = \rho_s V \lambda X \qquad (19)$$

where ρs is the particle density, V is the particle volume ($\pi D^3/6$), λ is the latent heat of water, D is the diameter of the particle, and X is the removed moisture. The drying time can then be evaluated from the ratio of Eqs. (19) and (18):

$$t = \frac{Q}{q} = \frac{\rho_s V \lambda X}{U_s A \Delta T_m} \qquad (20)$$

where U_s is approximate to $2K_f/D$, and K_f is the thermal conductivity of the gas film. Figures 9.17a,b present some additional examples of pneumatic dryers.

9.8 PACKAGING OF DEHYDRATED FOODS

The packaging of foods has the following functions (Karel, 1975; Paine and Paine, 1992): it is a material handling tool; processing aid; convenience item for the consumer; a marketing tool; cost-saving device; and protective device for the food. Thus, packaging should be regarded as an integral part of food processing (Brennan

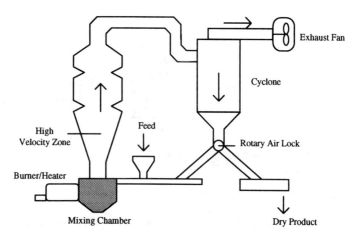

Figure 9.17a.   Rietz air lift pneumatic dryer. (Adapted from Williams-Gardner, 1971.)

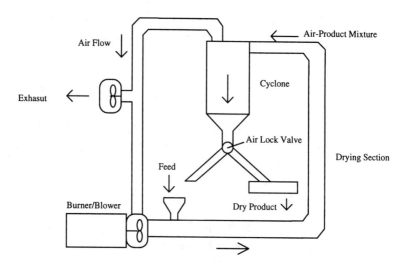

Figure 9.17b.   Proctor-Mark pneumatic dryer. (Adapted from Williams-Gardner, 1971.)

et al., 1990). In this section we discuss the protective role of packaging in maintaining the quality and extending the shelf life of dried foods.

The selection of a packaging material or container is based on the following factors (Karel, 1975; Eichner, 1986; Koszinowski and Priringer, 1986; De Leiris, 1986; Brennan et al., 1990).

Mechanical Damage

This type of damage results from the sudden impact or shock during handling and transport, as well as the vibration and compression loads imposed while the food is stored. It can be prevented by proper selection of strong, rigid packaging material and the inclusion of a cushioning material.

Permeability Characteristics

The loss of moisture results in less weight, as well as a deterioration in appearance and texture (i.e., meat and cheese). On the other hand, dried products tend to absorb moisture which can cause a loss of quality. Microbial spoilage or chemical deterioration occurs when water activity rises above a certain level. Gases and vapor can diffuse through materials by molecular diffusion as follows:

$$J = -DA\frac{dc}{dx} \tag{21}$$

where J is the flux of a particular gas, D is the diffusion coefficient, A is the area, c is the concentration, and x is the distance. The concentration term can be replaced by the solubility, S, and partial pressure, P, of the gas and Eq. (21) is:

$$J = -DSA\Delta P/\Delta x \tag{22}$$

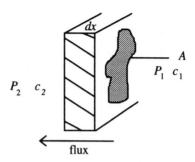

Figure 9.18.   Permeability scheme for a gas across a semipermeable membrane.

Figure 9.18 represents the system used to derive Eq. (22). The product *DS* is known as the permeability coefficient, *B*.

The permeability of some plastic films to oxygen at room temperature is listed in Table 9.2.

Temperature Change
Package performance and appearance must be retained when exposed to changes in temperature.

Light Transmission
Vitamin loss, color fading, and fat degradation are related to the exposure to light. The packaging must be opaque or

Table 9.2.   Oxygen permeabilities of plastic films.

Film	$B\left[\dfrac{cm^3 mm}{cm^2 s\, cm\, Hg}\right]$
Conventional polyethylene	6000–15000
High-density polyethylene	1500–3000
Pliofilm	200–5000
Cellophane	20–5000
Coated and waxed papers	100–15,000
Plastic laminations	10–400

From Brennan et al. (1990) and Paine and Paine (1992).

colored in order to exclude short lengthlight waves (Brennan et al., 1990). Based upon the Beer–Lambert law, the fraction of incident light transmitted through the packaging material can be expressed as:

$$I_x = I_o \exp(-\mu x) \tag{23}$$

where μ is the absorbance and x is a spatial coordinate. Then, the absorbance in a packaged product can be represented as:

$$\ln(I_o/I_x) = \mu_p x_p + \mu x \tag{24}$$

where μ_p is the absorbance of the packaging material and x_p is the thickness of the packaging material. Figure 9.19 presents the light absorption in a packaged food as discussed by Karel (1975).

Chemical and Biochemical Considerations
The packaging material must be compatible with the product with which it is in contact. Safety and quality

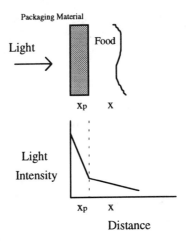

Figure 9.19.   Light absorption in packaged food. (Adapted from Karel, 1975.)

considerations must be taken into account when selecting a packaging material.

Microbiological and Biological Considerations

The prevention or reduction in microbial contamination is one of the functions of packages, as shown in Figure 9.20. The nature of packaging materials determines the protection offered by a package. Glass containers possessing the appropriate type of closure can be used in most packaging applications. Metal containers made from aluminum and tin-free steel are used mostly for liquid products. Paper and plastic materials are used as flexible packaging materials. This application includes the use of paper and plastics for the manufacture of wrapping, bags, envelopes, liners, and overwraps. Table 9.3 summarizes the general properties of films used for dried food packaging.

The packaging requirements for all dehydrated foods take into consideration two main causes of spoilage: moisture and oxygen. Dried foods are fragile, sensitive to light, and subject to cross contamination and insect attack (Sacharow and Griffin, 1970; Paine and Paine, 1992). Dried meat requires protection against moisture, mechan-

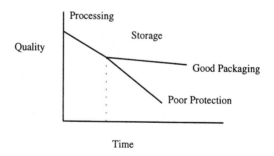

Figure 9.20. Quality loss during processing and storage. (Adapted from Karel, 1975.)

Table 9.3.   Properties and applications of films on dried foods.

Name	Permeability				Heat			
	Water	Gases	Volatiles	Oils	High T	Low T	Seal	Shrinkable
Regenerated cellulose								
Waxed coated	P	P	P	P	P	G	+	–
Copolymer coated	P	P	P	P	P	G	+	–
Polyethylene (PE)								
Low density	P	G	M	M	P	G	+	–
High density	P	M	M	M	M	G	+	–
Irradiated	M	M	M	M	M	G	+	+
Polypropylene								
Cast	P	P	P	P	G	P	+	–
Oriented	P	P	P	P	G	G	+	+/–
Rubber hydrochloride								
Pliofilm	V	V	G	M	P	V	+	+
Fluoroethylene	P	P	P	P	G	G	+	–

P = poor; M = moderate; V = variable; G = good, + = feasible; – not feasible.
Brennan et al., 1990; Paine and Paine, 1992.

ical damage and oxygen. Tin plate cans or heat-sealable laminates with aluminum foil are used in the packaging of dried meat products. Some examples of laminates are polyester–polyethylene–aluminum foil–polyethylene and cellophane–polyethylene–aluminum foil–polyethylene.

Poultry products should be either vacuum packaged or inert gas flushed and hermetically sealed in cans or pouches. Laminates for poultry products include cellophane–polyethylene– aluminum foil–polyethylene, cellophan– aluminum foil–polyamide, cellophan– aluminum foil–polyvinil chloride and rubber hydrochloride.

Dried egg whites are mainly protected from moisture absorption whereas whole eggs or yolks must also be protected from oxygen. This type of product is mainly sold in bulk packages. Vacuum or gas packed cans, canisters, sealed liners in carton boxes, or metal drums are used to package egg products.

Dried milk is packaged in aluminum foil laminates. Paper–polyethylene–foil–polyethylene is used in most dried milk applications. Large units are packaged in cans, jars, or in lined paperboard cartons.

Smoked or cured fish can be packaged in polyamide–polyethylene–polyester–polyethylene laminated films or high density polyethylene.

The major problem in the handling and storage of fruits and vegetables is insect attack. Packaging should be made from a material that insects will not penetrate. Wooden boxes, corrugated cartons, spiral-bound paper containers, and large tins are used for bulk packaging of fruits. Retail packaging includes paperboard folded cartons laminated with a liner or overwraps, coated cellophane bags, polyethylene, and polypropylene bags. Fruit flakes or powders are packaged in glass bottles, foil laminated flexible pouches, or in friction lidded tins. Dried legumes (i. e., beans, peas, lentils) are packaged in simple plastic bags of cellophane or low-density polyethylene films.

The usual gas for flushing dehydrated foods is nitrogen, which is inert with a low fat and moisture solubility. Carbon dioxide has also been used to modify packaging atmospheres (Fierheller, 1991).

9.9 CONCLUDING REMARKS

The main concepts for drying were used to develop additional techniques that reduce energy consumption, such as solar solar dryers, or improve product properties such as fluidized bed drying. Nevertheless, the best drying technique will be determined based on the type of product, its composition, and its physical properties.

Packaging of dried products is the next most important step after a drying operation. The selection of an appropriate packaging material plays an important role in the storage stability of the product. Again, the type of product, its composition, and physical properties have to be considered when selecting a packaging component.

9.10 REFERENCES

Arsem, H. B. and Ma, Y. H. 1990. Simulation of combined microwave and radiant freeze dryer. *Drying Technol.* 8(5):993–1016.

Bansal, N. K. and Garg, H. P. 1987. Solar crop drying. In *Advances in Drying,* Vol. 4, edited by A. S. Mujumdar. Hemisphere Publishing, New York.

Brennan, J. G., Butters, J. R., Cowell, N. D., and Lilley, A. E. U. 1990. *Food Engineering Operations,* Third edition. Elsevier Applied Science, New York.

Charm, S. E. 1978. Dehydration. In *The Fundamentals of Food Engineering,* Third edition. AVI Publishing, Westport, CT.

De Leiris, J. D. 1986. Water activity and permeability. In *Food Packaging and Preservation. Theory and practices,* edited by M. Mathlouthi. Elsevier Applied Science, New York.

Eichner, K. 1986. The influence of water content and water activity on chemical changes in foods of low moisture content under packaging aspects. In *Food Packaging and Preservation. Theory and practices,* edited by M. Mathlouthi. Elsevier Applied Science, New York.

Fierheller, M. G. 1991. Modified atmosphers packaging of micellaneous products. In *Modified Atmosphere Packaging of Food,* edited by B. Ooraikul and M. E. Stiles. Ellis Harwood Limited, Chichester, UK.

Garg, H. P. 1987. Solar food drying. In *Advances in Solar Energy Technology. Heating, Agricultural and Photovoltaic Applications of Solar Energy,* Vol. 3. D. Reidel Publishing, Dordrecht, Holland.

Geankoplis, C. J. 1983. Drying of Process Materials. In *Transport Processes and Unit Operations,* Second edition. Allyn and Bacon, Boston, MA.

Goswami, D. Y., Lavania, A., Shahbazi, S. and Masood, M. 1991. Analysis of a geodesic dome solar fruit dryer. *Drying Technol.* 9(3):677–691.

Harper, J. M. 1989. Food Extruders and Their Applications. In *Extrusion Cooking,* edited by C. Mercier, P. Linko and J. M. Harper. American Association of Cereal Chemists, Inc. St. Paul, MN.

Heldman, D. R. and Singh, R. P. 1981. Food dehydration. In *Food Process Engineering.* Second edition. AVI Publishing, New York.

Imre, L., Fábri, L., Gémes, L., and Hecker, G. 1990. Solar assisted dryer for seeds. *Drying Technol.* 8(2):343–349.

Karel, M. 1975. Protective packaging of foods. In *Principles of Food Science. Part II. Physical Principles of Food Preservation,* edited by M. Karel, O. R. Fennema, and D. B. Lund. Marcel Dekker, New York.

Koszinowski, J. and Piringer, O. 1986. Influence of the packaging material and the nature of the packed goods on the loss of volatile organic compounds. Permeation of aroma components. In *Food Packaging and Preservation.*

Theory and Practices, edited by M. Mathlouthi. Elsevier Applied Science, New York.

Masters, K. 1993. Importance of proper design of the air distributor plate in a fluidized bed system. *AIChE Symp. Series 297.* 89:118–126.

Norton, B. 1991. *Solar Energy Thermal Technology.* Springer-Verlag, London, UK.

Paine, F. A. and Paine, H. Y. 1992. *A Handbook of Food Packaging,* Second edition. Chapman & Hall, London, UK.

Rosselló, C., Berna, A., and Mulet, A. 1990. Solar drying of fruits in a mediterranean climate. *Drying Technol.* 8(2):305–321.

Sacharow, S. and Griffin, R. C. 1970. *Food Packaging.* AVI Publishing, Westport, CT.

Sochanski, J. S., Goyette, J., Bose, T. K., Akyel, C., and Bosisio, R. 1990. Freeze dehydration of foamed milk by microwaves. *Drying Technol.* 8(5): 1017–1037.

Spotts, M. R. and Waltrich, P. F. 1977. Vacuum dryers. *Chem. Eng.* 84(2): 120–123.

Stehli, D. and Escher, F. 1990. Design and continuous operation of a solar convection dryer with an auxiliary heating system. *Drying Technol.* 8(2): 241–260.

Tsamparlis, M. 1990. Solar drying for real applications. *Drying Technol.* 8(2): 261–285.

Williams-Gardner, A. 1971. *Industrial Drying.* Leonard Hill Books, London, UK.

APPENDICES

This section provides general information about physical properties of water and air.

Appendix 1.   Properties of superheated steam.

Absolute Pressure (kPa)		Temperature (°C)							
–Sat. Temp. (°C)		100	150	200	250	300	360	420	500
10–45.81	v	17.196	19.512	21.825	24.136	26.445	29.216	31.986	35.679
	h	2687.5	2783.0	2879.5	2977.3	3076.5	3197.6	3320.9	3489.1
	s	8.4479	8.6882	8.9038	9.1002	9.2813	9.4821	9.6682	9.8978
50–81.33	v	3.418	3.889	4.356	4.820	5.284	5.839	6.394	7.134
	h	2682.5	2780.1	2877.7	2976.0	3075.5	3196.8	3320.4	3488.7
	s	7.6947	7.9401	8.1580	8.3556	8.5373	8.7385	8.9249	9.1546
75–91.78	v	2.270	2.587	2.900	3.211	3.520	3.891	4.262	4.755
	h	2679.4	2778.2	2876.5	2975.2	3074.9	3196.4	3320.0	3488.4
	s	7.5009	7.7496	7.9690	8.1673	8.3493	8.5508	8.7374	8.9672
100–99.63	v	1.6958	1.9364	2.172	2.406	2.639	2.917	3.195	3.565
	h	2676.2	2776.4	2875.3	2974.3	3074.3	3195.9	3319.6	3488.1
	s	7.3614	7.6134	7.8343	8.0333	8.2158	8.4175	8.6042	8.8342
150–111.37	v		1.28	1.4443	1.6012	1.7570	1.9432	2.129	2.376
	h		2772.6	2872.9	2972.7	3073.1	3195.0	3318.9	3487.6
	s		7.4193	7.6433	7.8438	8.0720	8.2293	8.4163	8.6466
400–143.63	v		0.4708	0.5342	0.5951	0.6458	0.7257	0.7960	0.8893
	h		2752.8	2860.5	2964.2	3066.8	3190.3	3315.3	3484.9
	s		6.9299	7.1706	7.3789	7.5662	7.7712	7.9598	8.1913
700–164.97	v		0.2999	0.3363	0.3714	0.4126	0.4533	0.5070
	h		2844.8	2953.6	3059.1	3184.7	3310.9	3481.7
	s		6.8865	7.1053	7.2979	7.5063	7.6968	7.9299
1000–179.91	v		0.2060	0.2327	0.2579	0.2873	0.3162	0.3541
	h		2827.9	2942.6	3051.2	3178.9	3306.5	3478.5
	s		6.6940	6.9247	7.1229	7.3349	7.5275	7.7622
1500–198.32	v		0.1325	0.1519	0.1697	0.1899	0.2095	0.2352
	h		2796.8	2923.3	3037.6	3.1692	3299.1	3473.1
	s		6.4546	6.7090	6.9179	7.1363	7.3323	7.5698
2000–212.42	v			0.1114	0.1255	0.1411	0.1562	0.1757
	h			2902.5	3023.5	3159.3	3291.6	3467.6
	s			6.5453	6.7664	6.9917	7.1915	7.4317
2500–223.99	v			0.0870	0.0989	0.1119	0.1241	0.1399
	h			2880.1	3008.8	3149.1	3284.0	3462.1
	s			6.4085	6.6438	6.8767	7.0803	7.3234
3000–233.90	v			0.0706	0.0811	0.0923	0.1028	0.1162
	h			2855.8	2993.5	3138.7	3276.3	3456.5
	s			6.2872	6.5390	6.7801	6.9878	7.2338

v is the specific volume (m³/kg), *h* is the enthalpy (kJ/kg), and *s* is the entropy (kJ/kg K).

Appendix 2. Properties of saturated steam.

Temp.	Vapor pressure	Spec. Volume (m³/kg)		Enthalpy (kJ/kg)		Entropy (kJ/kg K)	
(°C)	(kPa)	Liq.	Sat. Vap.	H$_l$	H$_v$	Liq.	Sat. Vap.
0.01	0.611	0.0010002	206.14	0.00	2501.4	0.0000	9.1562
3	0.758	0.0010001	168.132	12.57	2506.9	0.0457	9.0773
6	0.935	0.0010001	137.734	25.20	2512.4	0.0912	9.0003
9	1.148	0.0010003	113.386	37.80	2517.9	0.1362	8.9253
12	1.402	0.0010005	93.784	50.41	2523.4	0.1806	8.8524
15	1.705	0.0010009	77.926	62.99	2528.9	0.2245	8.7814
18	2.064	0.0010014	65.038	75.58	2534.4	0.2679	8.7123
21	2.487	0.0010020	54.514	88.14	2539.9	0.3109	8.6450
24	2.985	0.0010027	45.883	100.70	2545.4	0.3534	8.5794
27	3.567	0.0010035	38.774	113.25	2550.8	0.3954	8.5156
30	4.246	0.0010043	32.894	125.79	2556.3	0.4369	8.4533
33	5.034	0.0010053	28.011	138.33	2561.7	0.4781	8.3927
36	5.947	0.0010063	23.940	150.86	2567.1	0.5188	8.3336
40	7.384	0.0010078	19.523	167.57	2574.3	0.5725	8.2570
45	9.593	0.0010099	15.258	188.45	2583.2	0.6387	8.1648
50	12.349	0.0010121	12.032	209.33	2592.1	0.7038	8.0763
55	15.758	0.0010146	9.568	230.23	2600.9	0.7679	7.9913
60	19.940	0.0010172	7.671	251.13	2609.6	0.8312	7.9096
65	25.03	0.0010199	6.197	272.06	2618.3	0.8935	7.8310
70	31.19	0.0010228	5.042	292.98	2626.8	0.9549	7.7553
75	38.58	0.0010259	4.131	313.93	2635.3	1.0155	7.6824
80	47.39	0.0010291	3.407	334.91	2643.7	1.0753	7.6122
85	57.83	0.0010325	2.828	355.90	2651.9	1.1343	7.5445
90	70.14	0.0010360	2.361	376.92	2660.1	1.1925	7.4791
95	84.55	0.0010397	1.982	397.96	2668.1	1.2500	7.4159
100	101.35	0.0010435	1.673	419.04	2676.1	1.3069	7.3549
105	120.82	0.0010475	1.419	440.15	2683.8	1.3630	7.2958
110	143.27	0.0010516	1.210	461.30	2691.5	1.4185	7.2387
115	169.06	0.0010559	1.037	482.48	2699.0	1.4734	7.1833
120	198.53	0.0010603	0.892	503.71	2706.3	1.5276	7.1296
125	232.1	0.0010649	0.771	524.99	2713.5	1.5813	7.0775
130	270.1	0.0010697	0.669	546.31	2720.5	1.6344	7.0269
135	313.0	0.0010746	0.582	567.69	2727.3	1.6870	6.9777
140	316.3	0.0010797	0.509	589.13	2733.9	1.7391	6.9299
145	415.4	0.0010850	0.446	610.63	2740.3	1.7907	6.8833

Temp.	Vapor pressure	Spec. Volume (m³/kg)		Enthalpy (kJ/kg)		Entropy (kJ/kg K)	
(°C)	(kPa)	Liq.	Sat. Vap.	H_l	H_v	Liq.	Sat. Vap.
150	475.8	0.0010905	0.393	632.20	2746.5	1.8418	6.8379
155	543.1	0.0010961	0.347	653.84	2752.4	1.8925	6.7935
160	617.8	0.0011020	0.307	675.55	2758.1	1.9427	6.7502
165	700.5	0.0011080	0.273	697.34	2763.5	1.9925	6.7078
170	791.7	0.0011143	0.243	719.21	2768.7	2.0419	6.6663
175	892.0	0.0011207	0.217	741.17	2773.6	2.0909	6.6256
180	1002.1	0.0011274	0.194	763.22	2778.2	2.1396	6.5857
190	1254.4	0.0011414	0.157	807.62	2786.4	2.2359	6.5079
200	1553.8	0.0011565	0.127	852.45	2793.2	2.3309	6.4323
225	2548	0.0011992	0.078	966.78	2803.3	2.5639	6.2503
250	3973	0.0012512	0.050	1085.36	2801.5	2.7927	6.0730
275	5942	0.0013168	0.033	1210.07	2785.0	3.0208	5.8938
300	8581	0.0010436	0.022	1344.0	2749.0	3.2534	5.7045

Appendix 3. Physical properties of water at the saturation pressure.

Temp. (°C)	Density (kg/m³)	β (x 10^{-4} K⁻¹)	c_p (kJ/kg K)	k (W/m K)	α (x 10^{-6} m²/s)	μ (x 10^{-6} Pa s)	ν (x 10^{-6} m²/s)
0	999.9	−0.7	4.226	0.558	0.131	1793.64	1.79
5	1000.0		4.206	0.568	0.135	1534.74	1.54
10	999.7	0.95	4.195	0.577	0.137	1296.44	1.30
15	999.1		4.187	0.587	0.141	1135.61	1.15
20	998.2	2.1	4.182	0.597	0.143	993.41	1.01
25	997.1		4.178	0.606	0.146	880.64	0.88
30	995.7	3.0	4.176	0.615	0.149	792.38	0.81
35	994.1		4.175	0.624	0.150	719.81	0.73
40	992.2	3.9	4.175	0.633	0.151	658.03	0.66
45	990.2		4.176	0.640	0.155	605.07	0.61
50	988.1	4.6	4.178	0.647	0.157	555.06	0.56
55	985.7		4.179	0.652	0.158	509.95	0.52
60	983.2	5.3	4.181	0.658	0.159	471.65	0.48
65	980.6		4.184	0.663	0.161	435.42	0.44
70	977.8	5.8	4.187	0.668	0.163	404.03	0.42
75	974.9		4.190	0.671	0.164	376.58	0.37
80	971.8	6.3	4.194	0.673	0.165	352.06	0.36
85	968.7		4.198	0.676	0.166	328.52	0.34
90	965.3	7.0	4.202	0.678	0.167	308.91	0.33
95	961.9		4.206	0.680	0.168	292.24	0.31
100	958.4	7.5	4.211	0.682	0.169	277.53	0.29
110	951.0	8.0	4.224	0.684	0.170	254.97	0.27
120	943.5	8.5	4.232	0.685	0.171	235.36	0.24
130	934.8	9.1	4.250	0.686	0.172	211.82	0.23
140	926.3	9.7	4.257	0.684	0.172	201.04	0.21
150	916.9	10.3	4.270	0.684	0.173	185.35	0.20
160	907.6	10.8	4.285	0.680	0.173	171.62	0.19
170	897.3	11.5	4.396	0.679	0.172	162.29	0.18
180	886.6	12.1	4.396	0.673	0.172	152.00	0.17
190	876.0	12.8	4.480	0.670	0.171	145.14	0.17
200	862.8	13.5	4.501	0.665	0.170	139.25	0.16
210	852.8	14.3	4.560	0.655	0.168	131.41	0.15
220	837.0	15.2	4.605	0.652	0.167	124.54	0.15
230	827.3	16.2	4.690	0.637	0.164	119.64	0.15
240	809.0	17.2	4.731	0.634	0.162	113.76	0.14
250	799.2	18.6	4.857	0.618	0.160	109.83	0.14

β is the coefficient of volumetric thermal expansion, c_p is the specific heat, k is the thermal conductivity, α is the thermal diffusivity, μ is the absolute viscosity, and ν is the kinematic viscosity.

Appendix 4. Physical properties of dry air at atmospheric pressure.

Temp. (°C)	Density (kg/m³)	β (x 10⁻³ K⁻¹)	c_p (kJ/kg K)	k (W/m K)	α (x 10⁻⁶ m²/s)	μ (x 10⁻⁶ Pa s)	ν (x 10⁻⁶ m²/s)
−20	1.365	3.97	1.005	0.0226	16.8	16.279	12.0
0	1.252	3.65	1.011	0.0237	19.2	17.456	13.9
10	1.206	3.53	1.010	0.0244	20.7	17.848	14.7
20	1.164	3.41	1.012	0.0251	22.0	18.240	15.7
30	1.127	3.30	1.013	0.0258	23.4	18.682	16.6
40	1.092	3.20	1.014	0.0265	24.8	19.123	17.6
50	1.057	3.10	1.016	0.0272	26.2	19.515	18.6
60	1.025	3.00	1.017	0.0279	27.6	19.907	19.4
70	0.996	2.91	1.018	0.0286	29.2	20.398	20.7
80	0.968	2.83	1.019	0.0293	30.6	20.790	21.5
90	0.942	2.76	1.021	0.0300	32.2	21.231	22.8
100	0.916	2.69	1.022	0.0307	33.6	21.673	23.6
120	0.870	2.55	1.025	0.0320	37.0	22.555	25.9
140	0.827	2.43	1.027	0.0333	40.0	23.340	28.2
160	0.789	2.31	1.030	0.0344	43.3	24.124	30.6
180	0.755	2.20	1.032	0.0357	47.0	24.909	33.0
200	0.723	2.11	1.035	0.0370	49.7	25.693	35.5
250	0.653	1.89	1.043	0.0400	60.0	27.557	42.2

β is the coefficient of volumetric thermal expansion, c_p is the specific heat, k is the thermal conductivity, α is the thermal diffusivity, μ is the absolute viscosity, and ν is the kinematic viscosity.

Appendix 5. Properties of ice.

Temperature (°C)	Thermal conductivity (W/m K)	Specific heat (kJ/kg K)	Density (kg/m³)
−101	3.50	1.382	925.8
−73	3.08	1.587	924.2
−45.5	2.72	1.783	922.6
−23	2.41	1.922	919.4
−18	2.37	1.955	919.4
−12	2.32	1.989	919.4
−7	2.27	2.022	917.8
0	2.22	2.050	916.2

INDEX